# MATHEMATICAL PROBLEMS RELATING TO THE NAVIER-STOKES EQUATION

## Series on Advances in Mathematics for Applied Sciences

**ISSN: 1793-0901**

*Published**:

Vol. 77 Continuum Thermodynamics
*by K. Wilmanski*

Vol. 78 Advanced Mathematical and Computational Tools in Metrology and Testing
*eds. F. Pavese et al.*

Vol. 79 From Genetics to Mathematics
*eds. M. Lachowicz and J. Miękisz*

Vol. 80 Inelasticity of Materials: An Engineering Approach and a Practical Guide
*by A. R. Srinivasa and S. M. Srinivasan*

Vol. 81 Stability Criteria for Fluid Flows
*by A. Georgescu and L. Palese*

Vol. 82 Applied and Industrial Mathematics in Italy III
*eds. E. De Bernardis, R. Spigler and V. Valente*

Vol. 83 Linear Inverse Problems: The Maximum Entropy Connection
*by H. Gzyl and Y. Velásquez*

Vol. 84 Advanced Mathematical and Computational Tools in Metrology and Texting IX
*eds. F. Pavese et al.*

Vol. 85 Continuum Thermodynamics
Part II: Applications and Examples
*by B. Albers and K. Wilmanski*

Vol. 86 Advanced Mathematical and Computational Tools in Metrology and Testing X
*eds. F. Pavese et al.*

Vol. 87 Mathematical Methods for the Natural and Engineering Sciences (2nd Edition)
*by R. E. Mickens*

Vol. 88 Road Vehicle Dynamics: Fundamentals of Modeling and Simulation
*by G. Genta and A. Genta*

Vol. 89 Advanced Mathematical and Computational Tools in Metrology and Testing XI
*eds. A. B. Forbes et al.*

Vol. 90 Advanced Mathematical and Computational Tools in Metrology and Testing XII
*eds. F. Pavese et al.*

Vol. 91 Fractal Analysis: Basic Concepts and Applications
*by C. Cattani, A. B. Mabrouk and S. Arfaoui*

Vol. 92 Chaos, Complexity, and Nonlinear Economic Theory
*by Wei-Bin Zhang*

*To view the complete list of the published volumes in the series, please visit:
https://www.worldscientific.com/series/samas

Series on Advances in Mathematics for Applied Sciences – Vol. 11

# MATHEMATICAL PROBLEMS RELATING TO THE NAVIER-STOKES EQUATION

Editor

**Giovanni Paolo Galdi**

*Istituto di Ingegneria*
*Università degli Studi di Ferrara*

World Scientific
*Singapore • New Jersey • London • Hong Kong*

*Published by*

World Scientific Publishing Co. Pte. Ltd.
5 Toh Tuck Link, Singapore 596224
*USA office:* 27 Warren Street, Suite 401-402, Hackensack, NJ 07601
*UK office:* 57 Shelton Street, Covent Garden, London WC2H 9HE

**British Library Cataloguing-in-Publication Data**
A catalogue record for this book is available from the British Library.

**Series on Advances in Mathematics for Applied Sciences — Vol. 11**
**MATHEMATICAL PROBLEMS RELATING TO THE NAVIER-STOKES EQUATIONS**



ISBN-13 978-981-02-0846-2
ISBN-10 981-02-0846-4
ISBN-13 978-981-02-3800-1 (pbk)
ISBN-10 981-02-3800-2 (pbk)

# FOREWORD

This volume collects four contributions to the mathematical theory of the Navier-Stokes equations. Specifically, the paper by C. Simader and H. Sohr deals with some basic questions related to one of the most fundamental tools in the theory, namely, the Helmholtz decomposition of the vector Lebesgue space $L^q$. The two works of Galdi are concerned with the classical topic of steady motion of a rigid body into a liquid with special emphasis to the study of the asymptotic behaviour at large distances of Leray's solutions. Finally, the contribution of Mogilevskii and Solonnikov treats the initial-boundary value problem in Hölder spaces in the presence of a free surface of liquid.

It is hoped that this volume will furnish new insights on classical problems and significant ideas on current topics as well and that it will contribute, in any case, to a better understanding of the fascinating theory of the Navier-Stokes equations.

The authors wish to thank the Deutsche Forschungsgemeinschaft (DFG), the Italian Consiglio Nazionale delle Ricerche (CNR) and MPI 40% and 60% contracts at the University of Ferrara for their financial support during the preparation of the book.

Ferraram, April 27th 1992 — *Giovanni Paolo Galdi*

# CONTENTS

# A new approach to the Helmholtz decomposition and the Neumann problem in $L^q$-spaces for bounded and exterior domains

Christian G. Simader
Universität Bayreuth, Mathematisches Institut,
D-8580 Bayreuth, Deutschland

and

Hermann Sohr
Universität Paderborn, Fachbereich Mathematik,
D-4790 Paderborn, Deutschland

## 1. Introduction

Let $G \subseteq \mathbb{R}^n$ be a domain with $n \geq 2$ and let $1 < q < \infty$. We consider the spaces

$$E^q(G) := \{\nabla p : p \in L^q_{loc}(\bar{G}),\ \nabla p \in L^q(G)^n\} \text{ equipped with norm } \|\nabla p\|_q$$

and

$$L^q_\sigma(G) := L^q\text{-closure of } C^\infty_{o,\sigma}(G) = \{u = (u_1,u_2,\dots,u_n) \in C^\infty_o(G)^n : \operatorname{div} u = 0\}$$

and study the problem of proving the direct decomposition

$$(1.1) \qquad L^q(G)^n = L^q_\sigma(G) \oplus E^q(G)$$

which is called the Helmholtz decomposition. Many thoughts have been spent on this problem since it is basic for the treatment of Navier-Stokes equations [2,4,6,7,10,16,17,18]. For bounded domains G, the most general result on this decomposition has been given by Fujiwara-Morimoto [2]. Their proof rests on the very general Lions-Magenes-theory for elliptic differential equations [8,9]. From the viewpoint of mathematical physics the case of exterior domains is more interesting than that of bounded domains. In the special case $n = 3$, Solonnikov [16] and Miyakawa [10] proved the decomposition of $(L^q(G))^n$ in exterior domains; Solonnikov used the integral equation method and Miyakawa applied similar tools as in [2]. In this case von Wahl [18] gave another proof of existence of weak solutions to the weak Neumann problem and of the Helmholtz decomposition *completely within the framework of integral equations in bounded and exterior domains*. To our knowledge, the case of n-dimensional exterior domains with $n \neq 3$, in particular the important two-dimensional case was an open problem until now except for $q = 2$; the case $q = 2$ is trivial by the Hilbert space method. It is the purpose of the present paper to give a rather elementary and self-contained new proof of the Helmholtz decomposition (1.1) in both bounded and exterior domains for general $1 < q < \infty$ and $n \geq 2$; in particular we include the case of two-dimensional exterior domains. We need only that the boundary $\partial G$ of $G$ is of class $C^1$ while in the above mentioned literature more regularity is necessary. Our proof of (1.1) rests on the following two basic theorems. The first one (Theorem 1.1) is well known by de Rham's much more general theory [12]; we give here a very elementary independent proof. For $q \neq 2$ the second basic tool is a variational inequality (see (1.3) below). If $\partial G$ is sufficiently smooth and $G$ is bounded, it can be derived from more general theories (e.g. Schechter [13]). But this method requires more regularity of the boundary. For Dirichlet's boundary value problem on bounded domains, similar estimates have been proved directly by Simader [14]. This inspired us to consider (1.3). In the contrary to the general theories [8], [9], [14] our new proof of this inequality is selfcontained and rather elementary, it includes also the case of exterior domains. For the whole space $G = \mathbb{R}^n$ the variational inequality (1.3) and in addition a type of regularity result (Lemma 3.3) is easily proved by means of a well known consequence (3.4) of the Calderon-Zygmund estimate. Moreover it turns out that inequalities (1.3) and (3.4) are equivalent (see Remark 3.4). The case of the half-space $H := \{x : x_n < 0\}$ is reduced to that of the whole space by a reflection argument (Lemma 3.6). A locally slighty bended half-space is considered as a small perturbation of the latter case (Lemma 3.8). Then a uniqueness result (Lemma 3.9) and at the end Theorem 1.3 are derived by means of a localization procedure. A further consequence of (1.3) is a functional representation (Theorem 1.3). Using this representation we can easily show the existence of a unique (up to an additive constant) weak solution $p \in L^q_{loc}(\bar{G})$ with $\nabla p \in L^q(\Omega)^n$ of Neumann's boundary value problem

$$(1.2) \qquad \Delta p = \operatorname{div} f \ , \ \langle N, (f-\nabla p)|_{\partial G}\rangle = 0,$$

where $f \in L^q(G)^n$ is given and $N$ is the exterior normal vector on $\partial G$. (Theorem 5.4). Thus we develop here really a new approach to the $L^q$-theory of (1.2) in the weak sense, and the Helmholtz decomposition can be considered as an application. In Theorem 4.2 we show that for $\nabla p \in E^q(G)$ with $\Delta p \in L^q(G)$ we have $\partial_N p|_{\partial G} = \langle N, \nabla p|_{\partial G}\rangle \in W^{-1/q,q}(\partial G)$. Conversely, given $g \in W^{-1/q,q}(\partial G)$ then

there exists $\nabla p \in E^q(G)$ with $\Delta p \in L^q(G)$ such that $g = \partial_N p|_{\partial G}$ (see Corollary 4.6). Via the Helmholtz decomposition those trace properties can be extended to vector fields. This admits the characterization of $L^q_\sigma(G)$ as those $L^q(G)^n$-fields u such that div u = 0 and $W^{-1/q,q}(\partial G) \ni u_N|_{\partial G} = 0$. Last property was a central point in [2]. We give here an easier proof (Theorems 4.4 and 5.2).
Conversely it turns out that the Helmholtz decomposition implies the variational inequality (1.3) (below) and the functional representation (Theorem 6.1).

Theorem 1.1. Let $n \geq 2$, $1 \leq q < \infty$ and let $G \subseteq \mathbb{R}^n$ be a domain. Let $u \in L^q_{loc}(G)^n$ be such that $\langle u,\phi\rangle = 0$ for all $\phi \in C^\infty_{o,\sigma}(G)$. Then there exists some $p \in H^{1,q}_{loc}(G)$ such that $u = \nabla p$ a.e. in G.

Corollary 1.2. Let the assumptions of Theorem 1.1 hold true and assume in addition that G is either a bounded or an exterior domain with boundary $\partial G$ of class $C^1$ and $u \in L^q(G)^n$. Then $u = \nabla p \in E^q(G)$.

Theorem 1.3. Let $n \geq 2$ and let $G \subseteq \mathbb{R}^n$ be a bounded or an exterior domain with boundary $\partial G$ of class $C^1$, let $1 < q < \infty$ and define $q'$ by $\frac{1}{q} + \frac{1}{q'} = 1$.

a) Then there exists a constant $C = C(G,q) > 0$ such that

$$(1.3)\quad \|\nabla p\|_q \leq C \sup_{0\neq\nabla\phi\in E^{q'}(G)} \frac{|\langle\nabla p,\nabla\phi\rangle|}{\|\nabla\phi\|_{q'}} \quad \text{for all } \nabla p \in E^q(G)$$

b) For $F^* \in (E^{q'}(G))^*$, $\|F^*\|_{(E^{q'})^*} := \sup_{0\neq\nabla\phi\in E^{q'}(G)} \frac{|F^*(\nabla\phi)|}{\|\nabla\phi\|_{q'}}$, there exists a unique $\nabla p \in E^q(G)$ such that $F^*(\nabla\phi) = \langle\nabla p,\nabla\phi\rangle$ for all $\nabla\phi \in E^{q'}(G)$ and
$$\|\nabla p\|_q \leq C\,\|F^*\|_{(E^{q'})^*} \leq C\|\nabla p\|_q$$
with the same constant $C > 0$ as in (1.3).

Some notations. If X is a Banach space, as usual by $X^*$ we denote the dual space equipped with norm $\|x^*\|_{X^*} := \sup_{0\neq x\in X} \frac{|\langle x^*,x\rangle|}{\|x\|_X}$, where $\langle x^*,x\rangle = x^*(x)$ means the value of $x^*$ at x. Throughout this paper $G \subseteq \mathbb{R}^n$ $(n \geq 2)$ denotes a domain, i.e. G is open and connected. G is called an exterior domain if G is a domain and there exists a bounded open set $K \neq \emptyset$ such that $G = \mathbb{R}^n\setminus\bar{K}$. $\bar{K}$ means the closure of K. Without loss of generality we may assume that $0 \in K$. We write $\partial G \in C^1$ if the boundary $\partial G$ is of class $C^1$. If A,B are subsets of $\mathbb{R}^n$ we write $A \subset B$ if A is contained in B or A equals B. We write $A \subset\subset B$ if A and B are

open, $\bar{A}$ compact and $\bar{A} \subset B$. If $x \in \mathbb{R}^n$ and $\sigma > 0, B_\sigma(x)$ denotes the open ball with radius $\sigma$ centered at $x$. If $M$ is a Lebesgue measurable subset of $\mathbb{R}^n$, $|M|$ denotes its Lebesgue measure.
Let $1 < q < \infty$ and let $q'$ be defined by $\frac{1}{q} + \frac{1}{q'} = 1$ that is $q' = \frac{q}{q-1}$.
Observe $(q')' = q$. $L^q(G)$ denotes the usual (real) Lebesgue space with norm $\|u\|_{L^q(G)} = \|u\|_q = (\int_G |u(x)|^q dx)^{1/q}$. By $\partial_i = \frac{\partial}{\partial x_i}$ $(i = 1,2,\dots,n)$ we denote the partial derivatives in the sense of distributions. Further $x = (x_1,x_2,\dots,x_n) \in G$ and $\nabla = (\partial_1,\partial_2,\dots,\partial_n)$ denotes the gradient, $\Delta = \partial_1^2 + \dots + \partial_n^2$ is the Laplacian and div $u = \partial_1 u_1 + \dots + \partial_n u_n$ is the divergence of $u = (u_1,u_2,\dots,u_n)$. $H^{1,q}(G)$ is the usual Sobolev space with norm $\|u\|_{H^{1,q}(G)} = \|u\|_q + \|\nabla u\|_q$ where $\|\nabla u\|_q = (\|\partial_1 u\|_q^q + \dots + \|\partial_n u\|_q^q)^{1/q}$. We need also the usual spaces $C^k(G)$ and $C_0^k(G)$ for $k = 0,1, \dots$ and for $k = \infty$. Here we distinguish between the usual spaces $L^q_{loc}(G)$ and the space $L^q_{loc}(\bar{G}) := \{u : u|_{G\cap B} \in L^q(G \cap B)$ for each ball $B$ with $G \cap B \neq \emptyset\}$, analogously for $H^{1,q}_{loc}(G)$ and $H^{1,q}_{loc}(\bar{G})$. In the same sense we use the notation $C_0^\infty(\bar{G}) := \{\phi|_{\bar{G}} : \phi \in C_0^\infty(\mathbb{R}^n)\}$. For bounded $G$ observe $C_0^\infty(\bar{G}) = C^\infty(\bar{G})$. The corresponding spaces of vector fields $u = (u_1,u_2,\dots,u_n)$ are denoted by $L^q(G)^n$, $L^q_{loc}(G)^n, \dots\dots$ The norm in $L^q(G)^n$ is again denoted by $\|u\|_{L^q(G)^n} =: \|u\|_{q,G} =: \|u\|_q$. We put $\langle u,v\rangle = \langle u,v\rangle_G = \int_G u_1 v_1 dx \dots + \int_G u_n v_n dx$ for vector fields $u = (u_1,\dots,u_n) \in L^q(G)^n$ and $v = (v_1,\dots,v_n) \in L^{q'}(G)^n$. We use the notation $\langle u,v\rangle = \langle u,v\rangle_G = \int_G u\, v\, dx$ also for scalar functions. Let $C^\infty_{0,\sigma}(G) := \{u \in C_0^\infty(G)^n : \text{div } u = 0\}$ and let $L^q_\sigma(G)$ be the $L^q(G)^n$-closure of $C^\infty_{0,\sigma}(G)$. We write $E^\infty(G) := \{\nabla v : v \in C_0^\infty(\bar{G})\}$. For $x \in \partial G$, $G \in C^1$ let $N = N(x)$ denote the exterior unit normal in $x$. The notation $u_N|_{\partial G} := \langle N,u|_{\partial G}\rangle$ has to be understood in the sense of the trace Theorem 5.3 (compare [2],p. 686) if $u \in L^q(G)^n$, div $u \in L^q(G)$. Throughout the paper $C$, $C_1$, $C_2, \dots$ are positive constants whose values may change from line to line. We make use of Friedrich's mollifier. Let $j \in C_0^\infty(\mathbb{R}^n)$, $0 \le j(z) \le 1$, $j(z) = j(-z)$, $j(z) = 0$ for $|z| \ge 1$ and $\int j(z)dz = 1$. Then we put $j_\epsilon(z) := \epsilon^{-n} j(\frac{z}{\epsilon})$ for $\epsilon > 0$ and $u_\epsilon(x) := \int_G j_\epsilon(x-z)u(z)dz$ for $u \in L^q(G)$ and we obtain $\|u-u_\epsilon\|_q \to 0$ as $\epsilon \to 0$.

Our main result on the Helmholtz decomposition reads as follows.

Theorem 1.4. (Main theorem). Let $n \geq 2$ and let $G \subseteq \mathbb{R}^n$ be a bounded domain or an exterior domain with boundary $\partial G$ of class $C^1$. Let $1 < q < \infty$. Then for every $f \in L^q(G)^n$ there exist uniquely determined $u \in L^q_\sigma(G)$ and $\nabla p \in E^q(G)$ such that the decomposition

(1.4) $\qquad f = u + \nabla p \qquad$ (Helmholtz decomposition)

is satisfied. Thus it holds $L^q(G)^n = L^q_\sigma(G) \oplus E^q(G)$ as a direct sum. $\nabla p$ and $f$ are linked by the equation $\langle \nabla p, \nabla \psi \rangle = \langle f, \nabla \psi \rangle$ for all $\nabla \psi \in E^{q'}(G)$. Further the estimates

(1.5) $\quad \|\nabla p\|_q \leq C\|f\|_q$ and $\|u\|_q \leq (C+1)\|f\|_q$ and $\|\nabla p\|_q + \|u\|_q \leq (2C+1)\,\|f\|_q$

with the same constant $C > 0$ as in (1.3) hold true. Setting $P_q f := u$ we obtain a bounded linear operator $P_q$ from $L^q(G)^n$ onto $L^q_\sigma(G)$ such that $P_q^2 = P_q$ (projection). It holds the density property

(1.6) $\quad L^q_\sigma(G) = \{u \in L^q(G)^n : \langle u, \nabla \phi \rangle = 0 \text{ for all } \nabla \phi \in E^{q'}(G)\}$.

Remark 1.5. If $q = 2$ we need no smoothness assumption on $\partial G$ if we replace $E^2(G)$ by the space $\tilde{E}^2(G) := \{\nabla p : p \in L^2_{loc}(G),\ \nabla p \in L^2(G)\}$ and (1.4) holds in the sense of an orthogonal decomposition. Theorem 1.4 is then an immediate consequence of Theorem 1.1. If $\partial G \in C^1$ then $\tilde{E}^2(G) = E^2(G)$ as is easily seen (compare e.g. [11], Théorèm 7.6, p. 114). In the general case $q \neq 2$ the proof of Theorem 1.4 rests on a combination of Theorems 1.1, 1.3 and Corollary 1.2. To our knowledge the result of Theorem 1.4 is new if $n \neq 3$, $q \neq 2$ and $G$ is an exterior domain.

*Acknowledgements. The authors would like to thank Professor Dr. Wolf von Wahl (Bayreuth) for many stimulating discussions on this and related topics.*

## 2. Proof of Theorem 1.1

Our elementary proof of Theorem 1.1 is based on the following two lemmas. Here we call a map $\gamma : [0,1] \to \mathbb{R}^n$ a piecewise $C^1$-curve if $\gamma$ is continuous and if there are points $0 = t_0 < t_1 < \ldots < t_k = 1$ such that the restriction *of $\gamma$ to every $[t_{i-1}, t_i]$ for $i = 1,\ldots,k$ is a $C^1$-map. We call $\gamma$ closed if* $\gamma(0) = \gamma(1)$, and we put $\bar{\gamma} := \{\gamma(t) : t \in [0,1]\}$. If $f \in C^0(\bar{\gamma})^n$ we put

$$\int_\gamma f ds := \int_0^1 f(\gamma(t))\cdot\gamma'(t)dt = \sum_{i=1}^{k}\int_{t_{i-1}}^{t_i} f(\gamma(t))\cdot\gamma'(t)dt \quad \text{where}$$

$$f(\gamma(t))\cdot\gamma'(t) = f_1(\gamma(t))\gamma_1'(t) + \ldots + f_n(\gamma(t))\gamma_n'(t) \quad ,$$

$$f = (f_1,\ldots,f_n), \quad \gamma = (\gamma_1,\ldots,\gamma_n).$$

Lemma 2.1. Let $n \geq 2$ and let $G \subseteq \mathbb{R}^n$ be a domain. Let $u \in L^1_{loc}(G)$ such that $\langle u,\phi\rangle = 0$ for all $\phi \in C^\infty_{o,\sigma}(G)$. Consider a closed piecewise $C^1$-curve $\gamma$ with $\bar{\gamma} \subset G$ and $0 < \epsilon < \epsilon_0 = \text{dist}(\bar{\gamma},\partial G)$ if $\partial G \neq \phi$ and $0 < \epsilon < \infty$ if $\partial G = \phi$. Then the mollified function $u_\epsilon$ satisfies $\int_\gamma u_\epsilon ds = 0$.

Proof. For $x \in G$, $0 < \epsilon < \epsilon_0$ we put $\phi_i^\epsilon(x) := \int_0^1 j_\epsilon(x-\gamma(t))\gamma_i'(t)dt$ $(i = 1,..,n)$, $\phi^\epsilon := (\phi_1^\epsilon,\ldots,\phi_n^\epsilon) \in C_o^\infty(G)^n$ and we get

$$\text{div}\,\phi^\epsilon(x) = \int_0^1 \sum_{i=1}^n (\partial_i j_\epsilon)(x-\gamma(t))\gamma_i'(t)dt = -\int_0^1 \frac{d}{dt} j_\epsilon(x-\gamma(t))dt = -j_\epsilon(x-\gamma(1))$$
$$+ j_\epsilon(x-\gamma(0)) = 0.$$

So $\phi^\epsilon \in C^\infty_{o,\sigma}(G)$ and therefore

$$0 = \langle u,\phi^\epsilon\rangle = \sum_{i=1}^n \int_G u_i(x)\int_0^1 j_\epsilon(x-\gamma(t))\gamma_i'(t)dtdx = \sum_{i=1}^n \int_0^1 \gamma_i'(t)\int_G j_\epsilon(x-\gamma(t))u_i(x)dxdt$$
$$= \sum_{i=1}^n \int_0^1 \gamma_i'(t)u_{i,\epsilon}(\gamma(t))dt = \int_\gamma u_\epsilon ds. \quad \square$$

Remark. Solenoidal fields like those in the proof above have been used in [2; p.697].

Lemma 2.2. Let $n \geq 2$ and let $G \subseteq \mathbb{R}^n$ be a domain. Let $1 \leq q < \infty$ and consider $p_i \in H^{1,q}_{loc}(G)$ for $i = 1,2,\ldots$ such that $(\nabla p_i)$ is a Cauchy-sequence in $L^q_{loc}(G)^n$. Then there is a sequence $(c_i)$ in $\mathbb{R}$ and some $p \in H^{1,q}_{loc}(G)$ such that

$(p_i - c_i)$ converges in $H^{1,q}_{loc}(G)$ to $p$. The sequence $(c_i)$ may be chosen independently of $q$.

Proof. i) For $x \in G$ let $d_x := (1/2)\text{dist}(x,\partial G)$ if $\partial G \neq \emptyset$ and $d_x := 1$ otherwise. Let $B(x) := \{y \in \mathbb{R}^n : |y-x| < d_x\}$ and denote by $|B(x)|$ the Lebesgue measure of $B(x)$. Then by the Poincaré inequality (compare e.g. [3], p. 157) there is a constant $C_x = C(|B(x)|) > 0$, independent of $q$, such that

$$\|f - |B(x)|^{-1}\int_{B(x)} f dy\|_{q,B(x)} \le C_x \|\nabla f\|_{q,B(x)} \text{ for } f \in H^{1,q}(B(x))$$

Let $c_i(x) := |B(x)|^{-1} \int_{B(x)} p_i dy$. Then we conclude from the last inequality and the hypothesis that $(p_i - c_i(x))$ is a Cauchy sequence in $L^q(B(x))$.

ii) Let $x,y \in G$ and let $H := B(x) \cap B(y) \neq \phi$.
Assume there are two sequences $(a_i)$, $(b_i)$ in $\mathbb{R}$ such that $(p_i - a_i)$ converges in $L^q(B(x))$ and $(p_i - b_i)$ converges in $L^q(B(y))$.

Since $|(a_i - b_i) - (a_k - b_k)| = |H|^{-\frac{1}{q}} \|(a_i - b_i) - (a_k - b_k)\|_{q,H}$

$\le |H|^{-\frac{1}{q}} \{\|(p_i - a_i) - (p_k - a_k)\|_{q,B(x)} + \|(p_i - b_i) - (p_k - b_k)\|_{q,B(y)}\}$.
$(a_i - b_i)$ converges in $\mathbb{R}$.

But then $p_i - a_i = (p_i - b_i) + (b_i - a_i)$ converges in $L^q(B(y))$ too.

iii) Choose now an arbitrary but fixed $x_o \in G$ and define $c_i := c_i(x_o)$.
Let $M := \{x \in G : (p_i - c_i)$ is a Cauchy sequence in $L^q(B(x))\}$. Then $x_o \in M$. $M$ is open: If $x \in M$ then every $y \in B(x)$ belongs to $M$ by definition of $B(y)$ and ii). $M$ is also closed in $G$: Let $(x_k) \subset M$, $x_k \longrightarrow z \in G$. For suitable $k_o \in \mathbb{N}$, $x_k \in B(z)$ for $k \ge k_o$. Then $B(x_{k_o}) \cap B(z) \neq \emptyset$ and by i) (applied to $B(z)$) and ii) the sequence $(p_i - c_i)$ is a Cauchy-sequence in $L^q(B(z))$. Since G is connected we conclude $M = G$.

iv) There is a sequence $(G_j)$ of open sets such that

$G_1 \subset\subset G_2 \subset\subset .. \subset\subset G$ and $G = \bigcup_{j=1}^{\infty} G_j$. Let $\hat{p}_j$ denote the limit of $(p_i - c_i)|_{G_j}$ in $H^{1,q}(G_j)$. Without loss of generality we may assume $\hat{p}_j|_{G_k} = \hat{p}_k$ for $j \ge k$. Then $p \in H^{1,q}_{loc}(G)$ is uniquely defined by $p(x) := \hat{p}_k(x)$ if $x \in G_k$ and $(p_i - c_i) \to p$ in $H^{1,q}_{loc}(G)$. □

<u>Proof of Theorem 1.1.</u> Choose a fixed $x_o \in G$ and denote for $i = 1,2,\ldots$
$G_i := \{x \in G$ : There is a piecewise $C^1$-curve $\gamma : [0,1] \longrightarrow G$ with $\gamma(0) = x_o$, $\gamma(1) = x$, $\frac{1}{i} < \text{dist}(\bar{\gamma}, \partial G)$ if $\partial G \neq \phi$ and $\sup[|\gamma(t)| : t \in [0,1]] < i\}$.
If $\partial G = \phi$, that is $G = \mathbb{R}^n$ put $G_i := B_i$. Clearly $G_i$ is open and connected and

$G_i \subset\subset G_{i+1}$. There is some $i_o$ such that $G_i \neq \phi$ for $i \geq i_o$; for simplicity we assume that $G_1 \neq \phi$. Then we obtain $G = \bigcup_{i=1}^{\infty} G_i$.
Let $d_i := \operatorname{dist}(G_{i+1}, \partial G_{i+2}) > 0$, choose $0 < \epsilon_i < \frac{d_i}{2}$ and consider the mollified functions $u_i = u_{\epsilon_i}$. If $\phi \in C^{\infty}_{o,\sigma}(G_{i+1})$ then $\phi_i = \phi_{\epsilon_i} \in C^{\infty}_{o,\sigma}(G_{i+2})$ and $<u_i,\phi> = <u,\phi_i> = 0$; $u_i$ is defined on $G_{i+1}$. If $x \in G_{i+1}$ then there is a piecewise $C^1$-curve $\gamma_x$ with $\gamma_x(0) = x_o$, $\gamma_x(1) = x$ and $\bar{\gamma}_x \subseteq G_{i+1}$. Assume that $\gamma^o_x$ is another curve of the same type and define

$\gamma^1(t) := \gamma_x(2t)$ for $t \in [0,1/2]$, $\gamma^1(t) := \gamma^o_x(2-2t)$ for $t \in [1/2,1]$.

Then $\overline{\gamma^1} \subseteq G_{i+1}$ and $\gamma^1$ is a closed piecewise $C^1$-curve. By Lemma 2.1 we get $0 = \int_{\gamma^1} u_i ds$. Since $\int_{\gamma^1} u_i ds = \int_{\gamma_x} u_i ds - \int_{\gamma^o_x} u_i ds$ we obtain by $p_i(x) = \int_{\gamma_x} u_i ds$, $x \in G_{i+1}$, a well defined function $\tilde{p}_i \in C^1(G_{i+1})$ with $\nabla \tilde{p}_i = u_i$ on $G_{i+1}$. We choose now a sequence $\varphi_i \in C^{\infty}_o(G)$, $i = 1,2,\ldots$, with $0 \leq \varphi_i \leq 1$, $\varphi_i = 1$ in a neighborhood of $G_i$, $\operatorname{supp} \varphi_i \subseteq G_{i+1}$. Put $p_i := \varphi_i \tilde{p}_i$. Then $p_i \in C^1(G)$ and if $i \geq k$ we get $\nabla p_i|_{G_k} = u_i$, $\|\nabla p_i - u\|_{L^q(G_k)} = \|u_i - u\|_{L^q(G_k)} \longrightarrow 0$ as $i \longrightarrow \infty$. Therefore $(\nabla p_i)$ converges to $u$ in $L^q_{loc}(G)^n$. By Lemma 2.2 the desired result follows. □

Proof of Corollary 1.2. Since $\partial G \in C^1$ and is compact and $\nabla p = u \in L^q(G)^n$, one readily sees $p \in L^q_{loc}(\bar{G})$ (compare e.g. [11], Théorème 7.6, p. 114) and by definition $\nabla p \in E^q(G)$. □

## 3. Proof of Theorem 1.3

In this section we give a very concentrated proof of Theorem 1.3 demonstrating the main ideas and the technical procedure. For more details and the corresponding estimates for the Dirichlet problem we refer to the forthcoming book [15]. First we study the functional analytic consequences of inequalities of type (1.3).

In the following let $G \subset \mathbb{R}^n$ be a domain and $1 < s < \infty$. We say that $G$ has the property $P_1(s)$ if

there exists a constant $C_s = C(s,G) > 0$ such that

$$(3.1.s)\qquad \|\nabla p\|_s \le C_s \sup_{0\ne\nabla\phi\in E^{s'}(G)} \frac{|<\nabla p,\nabla\phi>|}{\|\nabla\phi\|_{s'}}$$

holds for all $\nabla p \in E^s(G)$, where $s' = \frac{s}{s-1}$.

We say that G has the property $P_2(s)$ if

the map $\sigma_s : E^s(G) \to (E^{s'}(G))^*$ defined by $\sigma_s : \nabla p \longrightarrow <\nabla p,.>$ (that is $(\sigma_s(\nabla p))(\nabla\phi) = <\nabla p,\nabla\phi>$ for $\nabla p \in E^s(G)$ and $\nabla\phi \in E^{s'}(G)$) is a bijection and there is a constant $\tilde{C}_s = \tilde{C}(s,G) > 0$ such that for $\nabla p \in E^s(G)$

$$(3.2.s)\qquad \tilde{C}_s \|\nabla p\|_s \le \|\sigma_s(\nabla p)\|_{(E^{s'}(G))^*} \le \|\nabla p\|_s.$$

In our proof of Theorem 1.3 we make essential use of

<u>Lemma 3.1</u> Let $G \subset \mathbb{R}^n$ be a domain, $1 < q < \infty$ and $q' := \frac{q}{q-1}$. Then G has the property $P_1(s)$ for $s = q$ <u>and</u> $q'$ if and only if G has the property $P_2(s)$ for $s = q$ <u>and</u> $q'$.

<u>Proof.</u> We abbreviate $E^s := E^s(G)$. Observe $(s')' = s$. i) Suppose G has the property $P_1(s)$ for $s = q$ and $q'$. By (3.1.s) we conclude

$$(3.3.s)\qquad C_s^{-1} \|\nabla p\|_s \le \sup_{0\ne\nabla\phi\in E^{s'}} \frac{|<\nabla p,\nabla\phi>|}{\|\nabla\phi\|_{s'}} = \|\sigma_s(\nabla p)\|_{(E^{s'})^*} \le \|\nabla p\|_s$$

Therefore $\sigma_s(E^s)$ is a closed linear subspace of $(E^{s'})^*$. Suppose $\sigma_s(E^s) \subsetneqq (E^{s'})^*$. By the Hahn-Banach theorem there exists $F^{**} \in (E^{s'})^{**}$ such that $F^{**} \ne 0$ but $F^{**}|_{\sigma(E^s)} = 0$. Since $E^{s'}$ may be regarded as a closed subspace of the reflexive space $L^{s'}(G)^n$ it is reflexive too and we may identify $(E^{s'})^{**}$ with $E^{s'}$. Then there exists a unique $\nabla\phi \in E^{s'}$ such that $<F^{**},F^*> = F^*(\nabla\phi)$ for $F^* \in (E^{s'})^*$ and $\|\nabla\phi\|_{s'} = \|F^{**}\|_{(E^{s'})^{**}} > 0$. But for each $\nabla p \in E^s$ we then have $0 = (\sigma_s(\nabla p))(\nabla\phi) = <\nabla p,\nabla\phi>$ and therefore by (3.1.s') we conclude $\|\nabla\phi\|_{s'} = 0$ which is a contradiction. (3.2.s) follows from (3.3.s) with $\tilde{C}_s = C_s^{-1}$. ii) Suppose conversely that $P_2(s)$ holds for $s = q$ and $q'$. Then because $\sigma_{s'}(E^{s'}) = (E^s)^*$ and (3.2.s') we get for $\nabla p \in E^s$

$$\|\nabla p\|_s = \sup_{0\ne F^*\in(E^s)^*} \frac{F^*(\nabla p)}{\|F^*\|_{(E^s)^*}} \le \sup_{0\ne\nabla\phi\in E^{s'}} \frac{(\sigma_{s'}(\nabla\phi))(\nabla p)}{\tilde{C}_{s'}\|\nabla\phi\|_{s'}} = \tilde{C}_{s'}^{-1} \sup_{0\ne\nabla\phi\in E^{s'}} \frac{<\nabla p,\nabla\phi>}{\|\nabla\phi\|_{s'}}.$$

Therefore (3.1.s) holds with $C_s = \tilde{C}_{s'}^{-1}$. □

As an immediate consequence from Lemma 3.1 we get the following representation theorem.

Theorem 3.2 Let $G \subset \mathbb{R}^n$ be a domain and $1 < q < \infty$. Suppose G has the property $P_1(s)$ for $s = q$ and $q'$. Then for every $F \in (E^{s'}(G))^*$ there is a unique $\nabla p \in E^s(G)$ such that $\langle \nabla p, \nabla\phi\rangle = F(\nabla\phi)$ for all $\nabla\phi \in E^{s'}(G)$ ($s' = \frac{s}{s-1}$, $s = q$ or $s = q'$). Further there is a constant $C(s) > 0$ such that

$$C(s)^{-1}\|\nabla p\|_s \le \|F\|_s \quad (s = q \text{ or } q').$$

In the terminology above Theorem 1.3 tells that if G is a bounded or an exterior domain with $\partial G \in C^1$, then G has property $P_1(q)$ for all $1 < q < \infty$.

The proof of Theorem 1.3 is given in a number of steps, in fact we prove more than Theorem 1.3 says. First we show that the whole space and the half space have property $P_1(q)$ for $1 < q < \infty$. Then we proof that a sufficiently small "bended" half space has still property $P_1(q)$. Beside of the a-priori estimates we show in Lemmata 3.3 - 3.5 a type of "regularity property" decisively used later to prove uniqueness of the solution of the weak Neumann problem (Lemma 3.9). At the end the proof of Theorem 1.3 is reduced to the last cases by a usual localization procedure.

Lemma 3.3. Let $1 < q < \infty$, $1 < r < \infty$ and suppose $p \in L^r_{loc}(\mathbb{R}^n)$, $\nabla p \in L^r(\mathbb{R}^n)$ (that is $\nabla p \in E^r(\mathbb{R}^n)$) and

$$\sup_{0 \ne v \in C_0^\infty(\mathbb{R}^n)} \frac{|\langle \nabla p, \nabla v\rangle|}{\|\nabla v\|_{q'}} < \infty.$$

Then $\nabla p \in E^q(\mathbb{R}^n)$ and $\|\nabla p\|_q \le C_1 \sup_{0 \ne v \in C_0^\infty(\mathbb{R}^n)} \frac{|\langle \nabla p, \nabla v\rangle|}{\|\nabla v\|_{q'}}$

where $C_1 = C_1(n,q) > 0$.

Proof. Here we use the well known fact that $\{\Delta v : v \in C_0^\infty(\mathbb{R}^n)\}$ is dense in $L^\beta(\mathbb{R}^n)$ for all $1 < \beta < \infty$ and that the system $\nabla^2 v = (\partial_i\partial_j v)_{i,j=1,\dots,n}$ of all second order derivatives satisfies the estimate

$$(3.4) \qquad C_1\|\nabla^2 v\|_\beta \le \|\Delta v\|_\beta \text{ for } v \in C_0^\infty(\mathbb{R}^n)$$

with $C_1 = C_1(n,\beta) > 0$. For each fixed $i = 1,2,\dots,n$ we obtain

$$\sup_{0 \ne v \in C_0^\infty(\mathbb{R}^n)} \frac{|\langle \nabla p, \nabla v\rangle|}{\|\nabla v\|_{q'}} \ge \sup_{0 \ne w \in C_0^\infty(\mathbb{R}^n)} \frac{|\langle \nabla p, \nabla(\partial_i w)\rangle|}{\|\nabla(\partial_i w)\|_{q'}}$$

$$\geq \sup_{0 \neq w \in C_0^\infty(\mathbb{R}^n)} \frac{|<\partial_i p, \Delta w>|}{\|\nabla^2 w\|_{q'}} \geq C_1 \sup_{0 \neq w \in C_0^\infty(\mathbb{R}^n)} \frac{|<\partial_i p, \Delta w>|}{\|\Delta w\|_{q'}} .$$

From this we conclude that $v \longrightarrow <\partial_i p, v>$ is continuous on the dense subspace $M := \{\Delta w : w \in C_0^\infty(\mathbb{R}^n)\} \subset L^{q'}(\mathbb{R}^n)$ with respect to $\|.\|_q$. Therefore this functional may be uniquely and normpreserving extended to a continuous functional on the whole space $L^{q'}(\mathbb{R}^n)$. Therefore there is $g \in L^q(\mathbb{R}^n)$ such that $<\partial_i p, v> = <g, v>$ for all $v \in M$. From Weyl's lemma it follows that $W := \partial_i p - g$ is harmonic on $\mathbb{R}^n$. For fixed $x \in \mathbb{R}^n$ we represent $W$ by the second mean value property and get for $R > 0$

$$W(x) = |B_R(x)|^{-1}\Big( \int_{B_R(x)} \partial_i p(y)dy - \int_{B_R(x)} g(y)dy\Big) \quad \text{and}$$

by Hölder's inequality $|W(x)| \leq |B_R(x)|^{-1/r}\|\partial_i p\|_r + |B_R(x)|^{-1/q}\|g\|_q \to 0 \ (R \to \infty)$.

So $\partial_i p = g \in L^q(\mathbb{R}^n)$ and therefore

$$\sup_{0 \neq w \in C_0^\infty(\mathbb{R}^n)} \frac{|<\partial_i p, \Delta w>|}{\|\Delta w\|_{q'}} = \sup_{0 \neq f \in L^{q'}} \frac{|<\partial_i p, f>|}{\|f\|_{q'}} = \|\partial_i p\|_q .$$

Since $p \in L^1_{loc}(\mathbb{R}^n)$ we conclude from $\nabla p \in L^q(\mathbb{R}^n)$ that $p \in L^q_{loc}(\mathbb{R}^n)$, that is $\nabla p \in E^q(\mathbb{R}^n)$. □

Remark 3.4. In the proof of Lemma 3.3 we decisively used inequality (3.4) being a consequence of the Calderon-Zygmund theorem. Suppose conversely that the variational inequality $\|\nabla p\|_q \leq C \sup_{0 \neq v \in C_0^\infty(\mathbb{R}^n)} \frac{|<\nabla p, \nabla v>|}{\|\nabla v\|_{q'}}$ holds with a $C > 0$ for some $q$ with $1 < q < \infty$ and all $p \in C_0^\infty(\mathbb{R}^n)$. Then we conclude

$$\|\nabla \partial_i p\|_q \leq C \sup_{0 \neq v \in C_0^\infty(\mathbb{R}^n)} \frac{|<\nabla \partial_i p, \nabla v>|}{\|\nabla v\|_{q'}} = C \sup_{0 \neq v \in C_0^\infty(\mathbb{R}^n)} \frac{|<\Delta p, \partial_i v>|}{\|\nabla v\|_{q'}} \leq C \|\Delta p\|_q$$

which leads to (3.4) for $\beta = q$.

An immediate consequence is the following density property.

Corollary 3.5. Let $1 < q < \infty$. Then $E^\infty(\mathbb{R}^n) := \{\nabla v : v \in C_0^\infty(\mathbb{R}^n)\}$ is dense in $E^q(\mathbb{R}^n) = \{\nabla v : v \in L^q_{loc}(\mathbb{R}^n), \nabla v \in L^q(\mathbb{R}^n)^n\}$ with respect to the norm $\|\nabla v\|_q$.

Proof. Suppose $E^\infty(\mathbb{R}^n)$ is not dense in $E^q(\mathbb{R}^n)$. Then there exists $F^* \in (E^q(\mathbb{R}^n))^*$, $F^*|_{E^\infty} = 0$, $F^* \neq 0$. By Theorem 3.2 there exists a unique

$\nabla u \in E^{q'}(\mathbb{R}^n)$ with $\|\nabla u\|_{q'} > 0$ such that $F^*(\nabla p) = \langle\nabla u, \nabla p\rangle$ for all $\nabla p \in E^q(\mathbb{R}^n)$. But since $F^*(\nabla v) = 0$ for $\nabla v \in E^\infty(\mathbb{R}^n)$ by Lemma 3.3 we would conclude $\nabla u = 0$ contradicting $\|\nabla u\|_{q'} > 0$. □

Next we consider the half space $H = \{x = (x_1,..,x_n) \in \mathbb{R}^n: x_n < 0\}$. We denote $C_0^\infty(\bar{H}) := \{v|_H : v \in C_0^\infty(\mathbb{R}^n)\}$.

<u>Lemma 3.6.</u> Let $1 < q < \infty$, $1 < r < \infty$ and suppose $p \in L^r_{loc}(\bar{H})$, $\nabla p \in L^r(H)^n$ (that is, $\nabla p \in E^r(H)$) and

$$\sup_{0\neq v\in C_0^\infty(\bar{H})} \frac{|\langle\nabla p, \nabla v\rangle_H|}{\|\nabla v\|_{q',H}} < \infty .$$

Then $\nabla p \in E^q(H)$ and $\|\nabla p\|_{q,H} \le C_2 \sup_{0\neq v\in C_0^\infty(\bar{H})} \frac{|\langle\nabla p, \nabla v\rangle_H|}{\|\nabla v\|_{q',H}}$

where $C_2 = 2C_1 > 0$ with $C_1$ by Lemma 3.3.

<u>Proof.</u> We reduce this case to the previous one by a reflection argument. For any function $v : \mathbb{R}^n \to \mathbb{R}$ we put $\tilde{v}(x_1,x_2,\ldots,x_n) := v(x_1,x_2,\ldots,-x_n)$, and for any function $p : H \to \mathbb{R}$ we define the extension $p^* : \mathbb{R}^n \to \mathbb{R}$ by $p^*(x_1,\ldots,x_n) = p(x_1,\ldots,x_n)$ for $x_n < 0$ and $p^*(x_1,\ldots,x_{n-1},x_n) = p(x_1,\ldots,x_{n-1},-x_n)$ for $x_n \ge 0$. If $\nabla p \in E^r(H)$ then it is readily seen that $\nabla p^* \in E^r(\mathbb{R}^n)$, $(\partial_i p^*)(x) = (\partial_i p)(x)$ for $x_n < 0$ and $(\partial_i p^*)(x) = (-1)^{\delta_{in}} (\partial_i p)(x',-x_n)$ for $x_n > 0$. Further for $v \in C_0^\infty(\mathbb{R}^n)$ we get $\langle\nabla p^*, \nabla v\rangle_{\mathbb{R}^n} = \langle\nabla p, \nabla(v+\tilde{v})\rangle_H$ and $\|\nabla(v+\tilde{v})\|_{q',H} \le 2\|\nabla v\|_{q',\mathbb{R}^n}$ where $v+\tilde{v} \in C_0^\infty(\bar{H})$.

Then we get

$$\sup_{0\neq v\in C_0^\infty(\mathbb{R}^n)} \frac{|\langle\nabla p^*, \nabla v\rangle_{\mathbb{R}^n}|}{\|\nabla v\|_{q',\mathbb{R}^n}} \le 2 \sup_{v\in C_0^\infty(\mathbb{R}^n), v+\tilde{v}\neq 0} \frac{|\langle\nabla p, \nabla(v+\tilde{v})\rangle_H|}{\|\nabla(v+\tilde{v})\|_{q',H}} \le$$

$$\le 2 \sup_{0\neq w\in C_0^\infty(\bar{H})} \frac{|\langle\nabla p, \nabla w\rangle|}{\|\nabla w\|_{q',H}} .$$

By Lemma 3.3 we conclude $\nabla p^* \in E^q(\mathbb{R}^n)$ and since $p^*|_H = p$ and $\|\nabla p\|_{q,H} \le \|\nabla p^*\|_{q,\mathbb{R}^n}$ the estimate for $\|\nabla p\|_{q,H}$ is clear. □

In the next step we consider a "bended" half space. Let $\omega \in C^1(\mathbb{R}^{n-1})$ and put $x' = (x_1,x_2,\ldots,x_{n-1}) \in \mathbb{R}^{n-1}$. We suppose $\omega(x') = 0$ for all $|x'| \ge R$ where $R = R(\omega) > 0$. Then we define $H_\omega := \{x = (x',x_n) \in \mathbb{R}^n : x_n < \omega(x')\}$.

Lemma 3.7. Let $1 < q < \infty$ and let $\Omega$ denote either $H_\omega$ or a bounded or an exterior domain G with boundary $\nabla G \in C^1$. Then $E^\infty(\Omega) = \{\nabla v|_{\bar\Omega} : v \in C_0^\infty(\bar\Omega)\}$ is dense in $E^q(\Omega)$ with respect to $\|\nabla .\|_q$.

Proof: i) First we consider the case $\Omega = H_\omega$. Since $\partial H_\omega \in C^1$ and $\partial H_\omega \cap \{x : |x| > R\} = \partial H \cap \{x : |x| > R\}$ for $R = R(\omega)$, it is well known (see e.g. [1], p. 91) that for $\nabla v \in E^q(H_\omega)$ there exists an extension $\nabla\tilde v \in E^q(\mathbb{R}^n)$ such that $\nabla\tilde v|_{H_\omega} = \nabla v$. By Corollary 3.5 there is a sequence $(v_i)$ in $C_0^\infty(\mathbb{R}^n)$ such that $\|\nabla\tilde v - \nabla v_i\|_{q;\mathbb{R}^n} \to 0$. Then $\nabla v_i|_{H_\omega} \in E^\infty(H_\omega)$ and $\|\nabla v - \nabla v_i\|_{p,H_\omega} \le \|\nabla\tilde v - \nabla v_i\|_{q,\mathbb{R}^n} \to 0$.
ii) If $\Omega = G$ and G is either bounded or an exterior domain, then again since now $\partial G$ is compact like as in i) for given $\nabla v \in E^q(G)$ there is an extension $\nabla\tilde v \in E^q(\mathbb{R}^n)$. Then we proceed like as in i). □

Lemma 3.8. Let $1 < q < \infty$, $1 < r < \infty$. Then there is some $K = K(q,r,n) > 0$ with the following property: If $\|\nabla\omega\|_\infty = \sup_{x'\in\mathbb{R}^{n-1}}|\nabla\omega(x')| \le K$, then:

a) There are constants $C(s) = C(s,K,n) > 0$ such that

$$\|\nabla p\|_{s,H_\omega} \le C(s) \sup_{0\ne\nabla v\in E^\infty(H_\omega)} \frac{|\langle\nabla p,\nabla v\rangle|}{\|\nabla v\|_{s',H_\omega}}$$

holds for $s = q,q',r,r'$ and $\nabla p \in E^s(H_\omega)$.

b) If $\nabla p \in E^q(H_\omega)$ and $\sup_{0\ne\nabla v\in E^\infty(H_\omega)} \frac{|\langle\nabla p,\nabla v\rangle_{H_\omega}|}{\|\nabla v\|_{r',H_\omega}} < \infty$ then $\nabla p \in E^r(H_\omega)$.

Proof. i) First we prove a). For this purpose we introduce new coordinates which reduce $H_\omega$ to the half space H if $\|\nabla\omega\|_\infty$ is sufficiently small. We put $y_1(x) = x_1$, $y_2(x) = x_2$, ..., $y_{n-1}(x) = x_{n-1}$, $y_n(x) = x_n - \omega(x')$. Then $y \in C^1(\bar H_\omega)$, y maps $\bar H_\omega$ bijective on $\bar H$ and the Jacobian satisfies $J[y(x)] = 1$. Let $x = x(y)$ denote the inverse map and put $\tilde p(y) := p(x(y))$. Define $\tilde\partial_i := \partial/\partial y_i$, $\partial_i := \partial/\partial x_i$, $\nabla := (\partial_1,\dots,\partial_n)$, $\tilde\nabla := (\tilde\partial_1,\dots,\tilde\partial_n)$, $\omega_i := \partial\omega/\partial x_i$. Then an elementary calculation yields $\partial_i = \tilde\partial_i - \omega_i\tilde\partial_n$ for $i = 1,\dots,n-1$ and $\partial_n = \tilde\partial_n$ and for $\nabla p \in E^q(H_\omega)$ we see $\tilde\nabla\,\tilde p \in E^q(H)$.

Further with constants $d_i = d_i(s,n) > 0$ $(i = 1,2)$ independent of $\omega$ we get

$$\|\nabla v\|_{s',H_\omega} \le d_1(s')(1+\|\nabla\omega\|_\infty)\|\tilde\nabla\tilde v\|_{s',H} \text{ and}$$

$$|\langle\nabla p,\nabla v\rangle_{H_\omega}| \ge |\langle\tilde\nabla\tilde p,\tilde\nabla\tilde v\rangle_H| - d_2(s')\|\nabla\omega\|_\infty(1+\|\nabla\omega\|_\infty)\ \|\tilde\nabla\tilde p\|_{s,H}\|\tilde\nabla\tilde v\|_{s',H}$$

If we observe that the map $p \longmapsto \tilde p$ maps $E^s(H_\omega)$ onto $E^s(H)$ we derive by means of Lemma 3.6 (suppose $r = s$)

$$\sup_{0\neq\nabla v\in E^\infty(H_\omega)} \frac{|\langle\nabla p,\nabla v\rangle_{H_\omega}|}{\|\nabla v\|_{s',H_\omega}} \ge [d_1(s')(1+\|\nabla\omega\|_\infty)]^{-1}\left\{\sup_{0\neq\tilde\nabla\tilde v\in E^\infty(H)} \frac{|\langle\tilde\nabla\tilde p,\tilde\nabla\tilde v\rangle_H|}{\|\tilde\nabla\tilde v\|_{s',H}}\right.$$

$$\left. - d_2(s')\ d_1(s')^{-1}\right\}\|\nabla\omega\|_\infty\ \|\tilde\nabla\tilde p\|_{s,H}$$

$$\ge d_1(s')^{-1}[(1+\|\nabla\omega\|_\infty)^{-1}\ C_2(s)^{-1} - d_2(s')\|\nabla\omega\|_\infty]\ \|\tilde\nabla\tilde p\|_{s,H}$$

Choose $K \le 1$ such that $0 < K \le \min\{(4C_2(s)\ d_2(s'))^{-1} : s = q,q',r,r'\}$ with $C_2$ by Lemma 3.6. If $\|\nabla\omega\|_\infty \le K$, since $\|\nabla p\|_{s,H_\omega} \le d_1(s)(1+K)\ \|\tilde\nabla\tilde p\|_{s,H}$ we then get

$$\sup_{0\neq\nabla v\in E^\infty(H_\omega)} \frac{|\langle\nabla p,\nabla v\rangle_{H_\omega}|}{\|\nabla v\|_{s',H_\omega}} \ge C\ \|\nabla p\|_{s,H_\omega}$$

where $C(s) = [4\ C_2(s)\cdot d_1(s)\cdot d_1(s')(1+K)]^{-1}$ for $s = q,q',r,r'$, that is a).

ii) In order to prove b) let $\nabla p \in E^q(H_\omega)$. We consider first the case $q \ge r$. Here we use a localization procedure in order to reduce this case to the half space. Let $\varphi \in C^\infty(\mathbb{R}^n)$ such that $\varphi(x) = 0$ for $|x| \le R$, $\varphi(x) = 1$ for $|x| \ge 2R$ and $0 \le \varphi \le 1$; here $R = R(\omega) > 0$ denotes the constant with $\omega(x') = 0$ for $|x'| \ge R$. Let $L := \{x \in \mathbb{R}^n : x \in H,\ |x| < 2R\}$ We choose $h \in C_0^\infty(\bar H)$ and $c(h)\in\mathbb{R}$ such that $\int_L (h+c(h))dx = 0$. We put $v = \varphi\bar h$ where $\bar h = h+c(h)$ and by direct calculations we obtain

$$\langle\nabla p,\nabla v\rangle_{H_\omega} = \langle\nabla p,\nabla(\varphi\bar h)\rangle_{H_\omega} = \langle\nabla p,(\nabla\varphi)\bar h\rangle_H + \langle\nabla p,\varphi\nabla\bar h\rangle_H = \langle\nabla p,(\nabla\varphi)\bar h\rangle_H + \langle\nabla(\varphi p),\nabla\bar h\rangle_H$$

$$- \langle(\nabla\varphi)p,\nabla\bar h\rangle_H.$$

Using $\int_L \bar h\ dx = 0$ and Poincaré's inequality we obtain $\|\bar h\|_{r',L} \le c_1\|\nabla\bar h\|_{r',L}$ and $|\langle\nabla p,(\nabla\varphi)\bar h\rangle| \le c_2\|\nabla p\|_{r,L}\|\nabla\bar h\|_{r',L}$ and $\|\nabla(\varphi\bar h)\|_{r',L} = \|(\nabla\varphi)\bar h+\varphi(\nabla\bar h)\|_{r',L} \le c_3\|\nabla\bar h\|_{r',L}$, $|\langle(\nabla\varphi)p,\nabla\bar h\rangle| \le c_4\|p\|_{r,L}\ \|\nabla\bar h\|_{r',L}$. This leads to

$$\sup_{0\neq h\in C_o^\infty(\bar H)} \frac{|<\nabla(\varphi p),\nabla h>_H|}{\|\nabla h\|_{r',H}} = \sup_{0\neq h\in C_o^\infty(\bar H)} \frac{|<\nabla(\varphi p),\nabla \bar h>_H|}{\|\nabla \bar h\|_{r',H}}$$

$$\leq \Big[ \sup_{\substack{\bar h\in C^\infty(\bar H)\\ \varphi\bar h\neq 0}} \frac{|<\nabla p,\nabla(\varphi\bar h)>_H|}{\|\nabla(\varphi\bar h)\|_{r',H}} \Big] + c_5\|\nabla p\|_{r,L} + c_6\|p\|_{r,L}$$

$$\leq \Big[ \sup_{0\neq\nabla v\in E^\infty(H_\omega)} \frac{|<\nabla p,\nabla v>_{H_\omega}|}{\|\nabla v\|_{r',H_\omega}} \Big] + c_5\|\nabla p\|_{r,L} + c_6\|p\|_{r,L}.$$

From the hypothesis in b) we conclude since $q \geq r$ that the last expression is finite; thus we get from Lemma 3.6 that $\nabla(\varphi p) \in L^r(H)^n$. Since $q \geq r$ we conclude that $\nabla p \in L^r(H_\omega)^n$. Since $\partial H_\omega \in C^1$ we see like as in Corollary 1.2 that last property implies $\nabla p \in E^r(H_\omega)$.

iii) Next we consider the case $q < r$ in b). This case can be reduced to the previous one. Consider now $\nabla p \in E^q(H_\omega)$ and assume $q < r$. According to the assumption in b) by $F^*(\nabla v) := <\nabla p,\nabla v>$ for $\nabla v \in E^\infty(H_\omega)$ a continuous linear functional on the dense subspace $E^\infty(H_\omega)$ from $E^{r'}(H_\omega)$ is defined. By Theorem 3.2 the unique extension $\hat F^*$ of $F^*$ to the whole space $E^{r'}(H_\omega)$ may be represented with a uniquely determined $\nabla\hat p \in E^r(H_\omega)$ in the form $\hat F^*(\nabla v) = <\nabla\hat p,\nabla v>$ since by part i) of proof $H_\omega$ has properties $P_1(s)$ for $s = r, r', q, q'$. For $\nabla v \in E^\infty(H_\omega)$, $\hat F^*(\nabla v) = <\nabla\hat p,\nabla v> = <\nabla p,\nabla v>$ and since $\nabla p \in E^q(H_\omega)$ the restriction of $\hat F^*$ to $E^\infty(H_\omega)$ may be regarded as a functional continuous at $E^\infty(H_\omega)$ with respect to $\|\nabla .\|_{q'}$-norm. Since now $q < r$ we conclude by part ii) that $\nabla\hat p \in E^q(H_\omega)$. Then $<\nabla p-\nabla\hat p,\nabla v> = 0$ for $\nabla v \in E^\infty(H_\omega)$ and by density of $E^\infty(H_\omega)$ in $E^{q'}(H_\omega)$ and part a) for $s = q$ we conclude $\nabla p = \nabla\hat p$. But since $\nabla\hat p \in E^r(H_\omega)$ we end with $\nabla p \in E^r(H_\omega)$. □

Our proof of Theorem 1.3 uses decisively the following Lemma guaranteeing uniqueness (within the class $E^q(G)$) of the weak Neumann problem in $L^q$ (compare Corollary 4.3).

Lemma 3.9. Let $1 < q < \infty$, $n \geq 2$ and let $G \subsetneq \mathbb{R}^n$ be as in Theorem 1.4. Suppose $\nabla p \in E^q(G)$ and $<\nabla p,\nabla v> = 0$ for all $\nabla v \in E^{q'}(G)$. Then $\nabla p = 0$.

Proof. *i) We show that under the hypothesis of the lemma that $\nabla p \in E^2(G)$.* Since $<\nabla p,\nabla v> = 0$ especially for all $\nabla v \in E^\infty(G)$, by density of $E^\infty(G)$ in $E^2(G)$ we then conclude $<\nabla p,\nabla v> = 0$ for all $\nabla v \in E^2(G)$ and we may put $v = p$ and are finished. The property $\nabla p \in E^2(G)$ is clear if G is bounded and $q \geq 2$. In order to prove the remaining cases we show that every point $x_0 \in \bar{G}$ has an open neighborhood $U_{x_0}$ such that $\nabla p|_{G\cap U_{x_0}} \in L^2(G\cap U_{x_0})$. For exterior domains in addition we show that for a sufficiently big ball B cented at zero we have $\nabla p|_{G\setminus B} \in L^2(G\setminus B)$. This is done in the next steps.

ii) First let $x_0 \in \partial G$. After a translation we may assume $x_0 = 0$. Since $\partial G \in C^1$ there exists $\rho > 0$ and a function $\sigma \in C^1(\bar{B}_\rho)$ with $\nabla\sigma(0) \neq 0$ such that $G \cap B_\rho = \{x \in B_\rho : \sigma(x) < 0\}$ and $\partial G \cap B_\rho = \{x \in B_\rho : \sigma(x) = 0\}$ where $B_\rho := B_\rho(0)$. The most adequate parametrisation of $\partial G$ in a neighborhood of $x_0 = 0$ is found by projection on the tangential hyperplane at $x_0 = 0$. Essentially this is done in the following. Observe that $|\nabla\sigma(0)|^{-1}\,\nabla\sigma(0)$ equals the exterior unit normal of $\partial G$ at $x_0 = 0$. This procedure enables us by a suitable coordinate transform (see below) to come to the situation of Lemma 3.8.

There exists an orthogonal matrix S such that $S[\nabla\sigma(0)] = |\nabla\sigma(0)|e_n$, where $e_n = (\delta_{1n},\dots,\delta_{nn})$. Define $y(x) := Sx$ and put $\hat{\sigma}(y) := \sigma(S^{-1}y)$ for $y \in B_\rho$. Let

$G_\rho := G \cap B_\rho$, $\hat{G} := SG$ and $\hat{G}_\rho := \hat{G} \cap \hat{B}_\rho$. For $\nabla v \in E^s(G_\rho)$ we put $\hat{v}(y) := v(S^{-1}y)$ for $y \in \hat{G}_\rho$. Then $\nabla\hat{v} \in E^s(\hat{G}_\rho)$ and the norms $\|\nabla\hat{v}\|_{s,\hat{G}_\rho}$ and $\|\nabla v\|_{s,G_\rho}$ are equivalent. If $\nabla p \in E^s(G_\rho)$, $\nabla v \in E^{s'}(G_\rho)$ then $<\nabla\hat{p},\nabla\hat{v}>_{\hat{G}_\rho} = <\nabla p,\nabla v>_{G_\rho}$. We write $y = (y',y_n)$ with $y' \in \mathbb{R}^{n-1}$. Since $\nabla\hat{\sigma}(0) = |\nabla\sigma(0)|e_n \neq 0$ by the implicit function theorem we find $0 < \rho' < \rho$, $h > 0$ and a function $\psi \in C^1(\bar{B}'_{\rho'})$ where $B'_{\rho'} := \{y' \in \mathbb{R}^{n-1} : |y'| < \rho'\}$ with the following properties: Let $Z := Z_{\rho',h} = \{y \in \mathbb{R}^n : |y'| < \rho',\ |y_n| < h\}$ then $Z \subset \hat{B}_\rho$. For $y' \in B'_{\rho'}$ we have $(y',\psi(y')) \in Z$ and $\hat{\sigma}(y',\psi(y')) = 0$. Further $\psi(0) = 0$, $(\nabla'\psi)(0) = 0$ (where $\nabla' = (\partial_1,\dots,\partial_{n-1})$) and $\partial\hat{G} \cap Z = \{y \in Z : y_n = \psi(y')\}$, $\hat{G} \cap Z = \{y \in Z : y_n < \psi(y')\}$. Let $\eta \in C_0^\infty(\mathbb{R}^{n-1})$ such that $\eta(y') = 1$ for $|y'| \leq 1$ and $\eta(y') = 0$ for $|y'| \geq 2$. For $0 < \lambda < \rho'/2$ put $\eta_\lambda(y') := \eta(\lambda^{-1}y')$ and $\omega_\lambda(y') := \eta_\lambda(y')\,\psi(y')$ for $|y'| < \rho$ and $\omega_\lambda(y') := 0$ otherwise. We want to apply Lemma 3.8 iterated. For this purpose observe since $\psi(0) = |\nabla'\psi(0)| = 0$ we get $\sup\{|\nabla'\omega_\lambda(y')| : y' \in \mathbb{R}^{n-1}\} \to 0$ as $\lambda \to 0$. Let now $1 < q < 2 \leq n$ be given and denote by k the greatest integer smaller than $\frac{n}{q}$ and let $q_j := \frac{nq}{n-jq}$ for $j = 0,\dots,k$. Since $k + 1 \geq \frac{n}{q}$ we get $q_k \geq n \geq 2$. Let $K_0 = \min\{K(q_{j-1},q_j) : j = 1,\dots,k\}$ where $K(q_{j-1},q_j) > 0$ is the constant according to Lemma 3.8.

*Choose now* $\lambda > 0$ *so small that* $\|\nabla\omega_\lambda\|_\infty \leq K_0$ and let $H_{\omega_\lambda}$ be defined as above with respect to $\omega_\lambda$. Because of the equivalence of $\|\nabla\hat{v}\|_{s,\hat{G}_\rho}$ and $\|\nabla v\|_{s,G_\rho}$ and because of $<\nabla\hat{p},\nabla\hat{v}>_{\hat{G}_\rho} = <\nabla p,\nabla v>_{G_\rho}$ (reflecting the invariance of $\Delta$ against orthogonal transforms) we omit in the sequel the distinction between $\hat{v}$ and v, $\hat{G}$ and G etc. and assume that the translation $x_o \longmapsto 0$ and the above rotation of coordinates $x \longmapsto Sx$ are performed. Choose now $0 < r < \lambda$ such that $B_r \subset Z$ and let $r_j := r\cdot 2^{-(j+1)}$ for $j = 0,1,..,k$. Choose $\varphi_j \in C_o^\infty(B_{r_j})$ such that

$0 \le \varphi_j \le 1$ and $\varphi_j = 1$ on $B_{r_{j+1}}$. Let $G_j := G \cap B_{r_j}$.

Given $v \in C_o^\infty(\bar{H}_{\omega_\lambda})$, let $v_j := v - c_j(v)$ where $c_j(v) := |G_j|^{-1} \int_{G_j} v dx$.

Then $\varphi_j v_j$ is an admissible testing function and we get

$$0 = <\nabla p, \nabla(\varphi_j v_j)> = <\nabla(\varphi_j p), \nabla v_j> - <p\nabla\varphi_j, \nabla v_j> + <\nabla p, v_j \nabla\varphi_j> \text{ and therefore}$$

$$(3.5) \qquad <\nabla(\varphi_j p), \nabla v_j> = <p\nabla\varphi_j, \nabla v_j> - <\nabla p, v_j\nabla\varphi_j>.$$

We prove now by induction that $\nabla(\varphi_j p) \in E^{q_j}(H_{\omega_\lambda})$ for $j = 0,1,...,k$. The case $j = 0$ is clear. Let now $0 < j \le k$ and suppose $\nabla(\varphi_{j-1}p) \in E^{q_{j-1}}(H_{\omega_\lambda})$. Since $\varphi_{j-1} = 1$ on $G_j$, we conclude $\nabla p \in L^{q_{j-1}}(G_j)$ and by the Sobolev embedding theorem $p \in L^{q_j}(G_j)$ and $\|p\|_{q_j,G_j} \le d_{1,j}\|p\|_{H^{1,q_{j-1}}(G_j)}$.

Therefore with $M_j := \|\nabla\varphi_j\|_\infty$.

$$|<p\nabla\varphi_j, \nabla v_j>| \le M_j \|p\|_{q_j,G_j} \|\nabla v_j\|_{q_j',G_j} \le$$

$$\le d_{1,j} M_j \|p\|_{H^{1,q_{j-1}}(G_j)} \|\nabla v_j\|_{q_j'(H_{\omega_\lambda})}$$

$$|<\nabla p, v_j\nabla\varphi_j>| \le M_j \|\nabla p\|_{q_{j-1},G_j} \|v_j\|_{q'_{j-1},G_j}$$

Since $\frac{n}{q} + 1 - n < 1 \le j$ we conclude $q_j' = \frac{nq}{nq-n+jq} < n$ and by the Sobolev theorem $v_j \in L^{q_j'^*}(G_j)$ where

$$q_j'^* = \frac{nq_j'}{n-q_j'} = q'_{j-1} \text{ and } \|v_j\|_{q'_{j-1}} \le d_{2j}\|v_j\|_{H^{1,q_j'}(G_j)}.$$

Since $\int_{G_j} v_j dx = 0$ we get by the Poincaré inequality

$$\|v_j\|_{H^{1,q_j'}(G_j)} \le d_{3j} \|\nabla v_j\|_{q_j',G_j} \le d_{3j} \|\nabla v\|_{q_j',H_{\omega_\lambda}}.$$ Altogether we get

for $v \in C_o^\infty(\bar{H}_{\omega_\lambda})$

$$|<\nabla(\varphi_j p), \nabla v>| = |<\nabla(\varphi_j p), \nabla v_j>| \le d_{4j} \|p\|_{H^{1,q_{j-1}(G_j)}} \|\nabla v\|_{q_j',H_{\omega_\lambda}}.$$

By Lemma 3.8 b) we conclude $\nabla(\varphi_j p) \in E^{q_j}(H_{\omega_\lambda})$ and therefore

$\nabla p|_{G_{j+1}} \in L^{q_j}(G_{j+1})$.

iii) Suppose now that G is bounded and $1 < q < 2$. Then by step ii) of proof for every $x_o \in \partial G$ there is an open neighborhood $U_{x_o}$ such that $\nabla p|_{G\cap U_{x_o}} \in L^2(G \cap U_{x_o})$. $\partial G$ is covered by a finite number of those neighborhoods, say $U_{x_1},\ldots,U_{x_m}$. Then $G_1 := G \cap \bigcap_{i=1}^{m} (\mathbb{R}^N \setminus U_{x_i})$ is compact and contained in G. For each $x_o \in G_1$ there is $r = r(x_o) > 0$ such that $B_r(x_o) \subset\subset G$. With the same induction proof as in ii) using Lemma 3.3 instead of Lemma 3.8 we see $\nabla p|_{B_{r'}(x_o)} \in L^2(B_{r'}(x_o))$ where $r'(x_o) = r(x_o)\cdot 2^{-(k+1)}$. Since $G_1$ is covered by a finite number of those balls $B_{r'}$, we see $\nabla p|_{G_1} \in L^2(G_1)$ and altogether $\nabla p \in L^2(G)$.

iv) Let now G be an exterior domain with $C^1$-boundary, $G = \mathbb{R}^n \setminus K$ where $K \subset \mathbb{R}^n$ is compact. First we consider the case $1 < q < 2$. Choose $r > 0$ such that $K \subset B_r$ and put $r_j := (j+1)r$, $G_j := \mathbb{R}^n \setminus \bar{B}_{r_j}$, $j = 0,1,\ldots,k$, where k and $q_j$ are defined like as in part ii) of this proof. Choose $\psi_j \in C_o^\infty(B_{r_{j+1}})$, $0 \le \psi_j \le 1$ such that $\psi_j = 1$ in $B_{r_j}$ and put $\varphi_j := 1-\psi_j$. Then supp $\varphi_j \subset G_j$, $\varphi_j = 1$ in $G_{j+1}$. Observe supp $\nabla\varphi_j \subset R_j$ where $R_j = \{x \in \mathbb{R}^n: r_j < |x| < r_{j+1}\}$. Given $v \in C_o^\infty(\mathbb{R}^n)$, put $v_j := v - c_j(v)$ where $c_j(v) := |R_j|^{-1} \int_{R_j} v dx$ and consider (3.5) with the above meaning of $\varphi_j$ and $v_j$. We may now proceed like as in part ii) using Lemma 3.3 instead of Lemma 3.8 observing that because of supp$(\nabla\varphi_j) \subset R_j$ the integrals at the right hand side of (3.5) are taken over $R_j$. At the end we conclude that $\nabla(\varphi_k p) \in E^{q_k}(\mathbb{R}^n)$ where $q_k \ge 2$. This implies $\nabla p \in L^{q_k}(G_{k+1})$. Since $1 < q < 2 \le q_k$ we have with $G' := \{x \in G_{k+1} : |\nabla p(x)| \ge 1\}$ and $G'' := \{x \in G_{k+1} : |\nabla p(x)| < 1\}$

$$\int_{G_{k+1}} |\nabla p|^2 dx = \int_{G'} |\nabla p|^2 dx + \int_{G''} |\nabla p|^2 dx \le \int_{G'} |\nabla p|^{q_k} dx + \int_{G''} |\nabla p|^q dx < \infty$$

and therefore $\nabla p|_{G_{k+1}} \in L^2(G_{k+1})$. Finally like as in parts ii) and iii) we see $\nabla p|_{G\cap B_{k+1}} \in L^2(G_{k+1})$ and therefore $\nabla p \in L^2(G)$.

If $2 \le q < \infty$ we consider (3.5) with $\varphi_o$ and $v_o$ (see above). Since $p|_{R_o} \in H^{1,q}(R_o) \subset H^{1,2}(R_o)$, we immediately derive from (3.5) via the Poincaré inequality with a numerical constant $d_5 = d_5(R_o,q,n)$

$$|\langle\nabla(\varphi_o p),\nabla v\rangle| = |\langle\nabla(\varphi_o p),\nabla v_o\rangle| \le d_5 \, \|p\|_{H^{1,q}(R_o)} \, \|\nabla v\|_{2,\mathbb{R}^n}$$

By Lemma 3.3 we conclude $\nabla(\varphi_o p) \in E^2(\mathbb{R}^n)$ and therefore $\nabla p|_{G_1} \in L^2(G_1)$. Since

$G\backslash G_1$ is bounded and $q \geq 2$ we have in addition $\nabla p|_{G\backslash G_1} \in L^2(G\backslash G_1)$. So $\nabla p \in L^2(G)$ in this case too. □

Now we are in the position to prove (1.3).

Proof of Theorem 1.3. We first show that every point $x_0 \in \bar{G}$ (if G is exterior, then for $x_0 = \infty$ too) has an open neighborhood $U_{x_0}$ such that if p vanishes outside of $U_{x_0} \cap G$ an inequality of the desired type holds locally: For interior estimates like as considered in Lemma 3.3 and for estimates up to the boundary like as in Lemma 3.8. Then we proof Theorem 1.3 by contradiction via a partition of unity using the local estimates. The technical details are as follows.

i) Let $x_0 \in \partial G$. We proceed similar to part ii) of the proof of Lemma 3.9. In Lemma 3.8 let $r = q$ and consider the constant $K = K(q) > 0$. Like as in the proof of Lemma 3.9 after a rotation of coordinates choose $\lambda > 0$ so small that $\|\nabla\omega_\lambda\|_\infty \leq K$. Choose $r = r(x_0)$ with $0 < r < \lambda$ such that $B_r(x_0) \subset Z$ and choose a function $\varphi \in C_0^\infty(B_r(x_0))$, $0 \leq \varphi \leq 1$, $\varphi(x) = 1$ for $x \in B'(x_0) := B_{r/2}(x_0)$. If $\nabla p \in E^q(G)$, then $\nabla(\varphi p) \in E^q(G)$ too and outside of $G \cap B_r(x_0)$ we may continue $\varphi p$ by zero to the whole space $H_{\omega_\lambda}$. By Lemma 3.8 there is $C_3(q) > 0$ such that

$$(3.6) \qquad \|\nabla(\varphi p)\|_q \leq C_3 \sup_{0 \neq \nabla v \in E^\infty(H_{\omega_\lambda})} \frac{|\langle\nabla(\varphi p)\nabla v\rangle|}{\|\nabla v\|_{q',H_{\omega_\lambda}}}$$

Since $\partial G$ is compact we find a finite number of points $x_i (i = 1,\ldots,M)$, $r_i = r_i(x_i) > 0$, balls $B_i' := B_{r_i/2}(x_i) \subset B_{r_i}(x_i) := B_i$, $\varphi_i \in C_0^\infty(B_i), 0 \leq \varphi_i \leq 1$, $\varphi_i(x) = 1$ for $x \in B_i'$ and $\omega_i := \omega_{\lambda_i}^{(i)}$ such that (3.6) holds true with $\varphi$ replaced by $\varphi_i$ and $H_{\omega_\lambda}$ by $H_{\omega_i}$. Finally $\partial G \subset \bigcup_{i=1}^{M} B_i'$.

ii) If G is bounded, $G_1 := G \cap \bigcap_{i=1}^{M} (\mathbb{R}^N\backslash B_i')$ is compact and $G_1 \subset G$. Since $r := \operatorname{dist}(G_1,\partial G) > 0$, we find $x_i \in G_1$ $(i = M+1,..,P)$ such that $G_1 \subset \bigcup_{i=M+1}^{P} B_i'$, where $B_i' := B_{r/2}(x_i)$ and $B_i := B_r(x_i) \subset G$. Choose $\varphi_i \in C_0^\infty(B_i)$, $\varphi_i = 1$ in $B_i'$.

$$(3.7) \qquad \|\nabla(\varphi_i p)\|_q \leq C_i^i \sup_{0 \neq v \in C_0^\infty(\mathbb{R}^n)} \frac{|\langle\nabla(\varphi_i p),\nabla v\rangle|}{\|\nabla v\|_{q',\mathbb{R}^n}}$$

iii) If $K \subset \mathbb{R}^n$ is compact, $G = \mathbb{R}^n\backslash K$ an exterior domain with $\partial G \in G^1$. Then choose $r > 0$ such that $K \subset B_r$. Choose $\psi \in C_o^\infty(B_{2r})$, $0 \le \psi \le 1$, $\psi = 1$ in $B_r$ and put $\varphi_o := 1-\psi$. then $\operatorname{supp} \nabla\varphi_o \subset R_o := \{x \in \mathbb{R}^n : r \le |x| \le 2r\}$. If $\nabla p \in E^q(G)$, then $\nabla(\varphi_o p) \in E^q(\mathbb{R}^n)$ and (3.7) holds for $\varphi_o p$. Now $G_{2r} := G \cap B_{2r}$ is bounded and may be covered as in i) or ii) respectively by finitely many $B_i'$, $i = 1,..,P$

iv) Suppose that the statement of Theorem 1.3 is not true. Then there is a sequence $(\nabla p_k) \subset E^q(G)$ such that $\|\nabla p_k\|_q = 1$ and $\sup\limits_{0 \ne \nabla v \in E^{q'}(G)} \dfrac{|\langle\nabla p_k, \nabla v\rangle|}{\|\nabla v\|_{q'}} =: \epsilon_k \to 0.$

Without loss of generality we may assume $\int_G p_k dx = 0$ if G is bounded and $\int_{G_{2r}} p_k dx = 0$ if G is an exterior domain. Since $E^q(G)$ is reflexive there is $\nabla p \in E^q(G)$ and a subsequence (again denoted by $p_k$) such that $\nabla p_k$ converge weakly to $\nabla p$. For $\nabla v \in E^{q'}(G)$ we then get $\langle\nabla p, \nabla v\rangle = \lim\limits_{k\to\infty} \langle\nabla p_k, \nabla v\rangle = 0$ and by Lemma 3.9 $\nabla p = 0$. By the Poincaré inequality we get $\|p_k\|_{H^{1,q}(G)} \le$ const for G bounded and $\|p_k\|_{H^{1,q}(G_{2r})} \le$ const for G exterior domain. In both cases we conclude by Rellich's theorem that $p_k \to p$ strongly in $L^q(G)$ resp. $L^q(G_{2r})$. Then we get $p = 0$ that is $p_k \to 0$ in $L^q(G)$ resp. $L^q(G_{2r})$. Fix now any $i \in \{0,..,M,M+1,..P\}$. If $i \in \{1,..M\}$ let $\Omega := H_{\omega_i}$, if $i = 0$ or $i = m+1,..,P$ let $\Omega := \mathbb{R}^n$. For each fixed i we then have by (3.6) and (3.7) with $C_i > 0$

$$(3.8) \qquad C_i \|\nabla(\varphi_i p_k)\|_q \le \sup_{0 \ne \nabla v \in E^{q'}(\Omega)} \frac{\langle\nabla(\varphi_i p_k), \nabla v\rangle}{\|\nabla v\|_{q'}} =: d_k$$

For each $k \in \mathbb{N}$ there is $\nabla v_k \in E^{q'}(\Omega)$ with $\|\nabla v_k\|_{q'} = 1$ and $0 \le d_k - \langle\nabla(\varphi_i p_k), \nabla v_k\rangle \le \frac{1}{k}$. For $i = 1,..,M,M+1,..,P$ we may assume $\int_{\Omega\cap B_i} v_k dx = 0$ and for $i = 0$ : $\int_{R_o} v_k dx = 0$(for $R_o$ compare part iii)). So by the Poincaré inequality we conclude $\|v_k\|_{H^{1,q'}(B_i\cap\Omega)} \le$ const and $\|v_k\|_{H^{1,q'}(R_o)} \le$ const respectively. We select a subsequence again denoted by $(v_k)$ and $\nabla v \in E^{q'}(\Omega)$ such that $\nabla v_k$ converges weakly to $\nabla v$ in $E^{q'}(\Omega)$ and $v_k \to v$ in $L^{q'}(B_i\cap\Omega)$ resp. $L^{q'}(R_o)$ strongly. Then

$$\begin{aligned} d_k &\le \frac{1}{k} + \langle\nabla(\varphi_i p_k), \nabla v_k\rangle \\ &= \frac{1}{k} + \langle\nabla p_k, \nabla(\varphi_i v_k)\rangle + \langle p_k \nabla\varphi_i, \nabla v_k\rangle - \langle\nabla p_k, v_k \nabla\varphi_i\rangle \end{aligned}$$

$$\le \frac{1}{k} + \epsilon_k \ \|\nabla(\varphi_i v_k)\|_{q'} + |\langle p_k \nabla\varphi_i, \nabla v_k\rangle| + |\langle \nabla p_k, v_k \nabla\varphi_i\rangle|$$

Since at the support of $\nabla\varphi_i$ we have $p_k \to 0$ in $L^q$-norm we conclude $\langle p_k\nabla\varphi_i, \nabla v_k\rangle \to 0$. Analogously $v_k \to v$ in $L^q$-norm at the support of $\nabla\varphi_i$ and $\nabla p_k \underset{w}{\longrightarrow} 0$ weakly which gives $|\langle\nabla p_k, v_k\nabla\varphi_i\rangle| \to 0$. Since $\epsilon_k \to 0$ and $\|\nabla(\varphi_i v_k)\|_{q'} \le$ const we get $d_k \to 0$ and by (3.8) $\|\nabla(\varphi_i p_k)\|_q \to 0$ that is $\|\nabla p_k\|_{q,B_i'} \to 0 (i = 1,..,P)$ if G is bounded resp. $\|\nabla p_k\|_{q,G\backslash B_{2r}} \to 0$ if $i = 0$ (if G is exterior domain). Since $G \subset \bigcup_{i=1}^{P} B_i'$ and $G \subset G\backslash B_{2r} \cup \bigcup_{i=1}^{P} B_i'$ if G is an exterior domain we get $\|\nabla p_k\|_{q,G} \to 0$ contradicting $\|\nabla p_k\|_{q,G} = 1$. So we proved (1.3). The second part of Theorem 1.3 is a consequence of the first part via Theorem 3.2. □

## 4. Traces.

Let G be bounded or an exterior domain with boundary $\partial G \in C^1$ and define for $1 < q < \infty$

$$F^q(G) := \{\nabla p \in E^q(G) : \Delta p \in L^q(G)\}$$

and

$$\|\nabla p\|_{F^q} := (\|\nabla p\|_q^q + \|\Delta p\|_q^q)^{\frac{1}{q}} \quad \text{for} \quad \nabla p \in F^q(G)$$

We show that for $\nabla p \in F^q$ there exists a uniquely defined trace $\partial_N p|_{\partial G} := \langle N, \nabla p|_{\partial G}\rangle$ in the sense of distributions. First we prove a density property. We use similar arguments as in [17], p. 6-7. The case of an exterior domain demands an additional consideration.

Lemma 4.1. Let G be either bounded or an exterior domain with boundary $\partial G \in C^1$. Then for $1 < q < \infty$, $E^\infty(G)$ is dense in $F^q(G)$ with respect to the $F^q$-norm.

Proof. i) If $\varphi \in C_0^\infty(\bar G)$ and $\nabla p \in F^q(G)$ then because of $p \in L^q_{loc}(\bar G)$ we get $\nabla(p\varphi) \in F^q(G)$ and $\nabla(p\varphi) = \varphi\nabla p + p\nabla\varphi$, $\Delta(p\varphi) = \varphi\Delta p + 2\nabla\varphi\nabla p + p\Delta\varphi \in L^q(G)$ in the sense of distributions.

ii) Let G be bounded. If $x_o \in \partial G \in C^1$ then eventually after renumbering coordinates there exist $r = r(x_o) > 0$, $h = h(x_o) > 0$ and $\omega \in C^1(\overline{B_r'(x_o)})$ where $B_r'(x_o) := \{x' \in \mathbb{R}^{N-1}: |x'-x_o'| < r\}$ (write $x = (x', x_n)$, $x' \in \mathbb{R}^{n-1}$) with the following property:

Let $Z_{r,h}(x_o) := \{(x', x_n) \in \mathbb{R}^n : x' \in B_r'(x_o), |x_n - \omega(x')| < h\}$.

Then $G \cap Z_{r,h}(x_o) = \{x \in Z_{r,h} : \omega(x') - h < x_n < \omega(x')\}$ and $\partial G \cap Z_{r,h}(x_o) = \{x \in Z_{r,h} : x_n = \omega(x')\}$. Then $\partial G$ may be covered by a finite number $\tilde{Z}_i := Z_{r_i/4,h_i/4}(x_i)(i=1,..,M)$ of such sets. Since $G_1 := G \cap \bigcap_{i=1}^{M}(\mathbb{R}^n \setminus \tilde{Z}_i)$ is compact and $G_1 \subset G$ there are finitely many balls $B_i \subset\subset G$ $(i = M+1,..,P)$ covering $G_1$. Choose a partition of unity $0 \le \varphi_i \in C_o^\infty(\tilde{Z}_i)$ or $\varphi_i \in C_o^\infty(B_i)$ respectively subordinate to this covering. If $p \in F^q(G)$ then $p = \sum_{i=1}^{P} p_i$ where $p_i = \varphi_i p$, $\nabla p_i \in F^q(G)$, supp $p_i \subset \tilde{Z}_i$ or $B_i$ respectively. So it suffices to approximate each $p_i$ by $E^\infty(G)$ functions. If supp $p_i \subset \tilde{Z}_i$ and $0 < \delta < h_i/8$ put $p_i^\delta(x) := p_i(x-\delta e_n)$ for $x \in Z_{r_i,h_i}$. Then clearly $p_i^\delta \in F^q(G)$ and $\|\nabla p_i - \nabla p_i^\delta\|_{F^q} \to 0$ as $\delta \to 0$. Let $K := \|\nabla\omega\|_{\infty,B'_{r_i}(x_i)}$ and $0 < \epsilon < \min(h_i/4, r_i/4, \delta(1+K)^{-1})$ and put $p_{i\epsilon}^\delta(x) := \int j_\epsilon(x-y)p_i^\delta(y)dy = \int_G j_\epsilon(x-\delta e_n - z)p_i(z)dz$. Then $(p_i^\delta)_\epsilon(x) = 0$ for $x \in Z_{r_i,h_i} \setminus Z_{r_i/2,h_i/2}$.
For $x \in Z_{r_i/2,h_i/2}(x_i) \cap G$ and $|z-(x-\delta e_n)| \le \epsilon$ we get $z \in Z_{r_i,h_i}(x_i) \cap G$. Therefore the function $\phi(z) := j_\epsilon(x-\delta e_n - z)$ has compact support in $Z_{r_i,h_i}(x_i) \cap G$ and then $(\nabla p_i{}^\delta{}_\epsilon)(x) = (\nabla p_i{}^\delta)_\epsilon(x)$ and $\Delta p_{i\epsilon}^\delta(x) = (\Delta p_i{}^\delta)_\epsilon(x)$. So $p_i{}^\delta{}_\epsilon \in C^\infty(\bar{G})$ and $\|\nabla p_i{}^\delta - \nabla p_i{}^\delta{}_\epsilon\|_{F^q} \to 0$ $(\epsilon \to 0)$. If supp $p_i \subset B_i$ then $p_{i\epsilon} := j_\epsilon * p_i \in C_o^\infty(G)$ for $0 < \epsilon < \text{dist}(\overline{B_i}, \partial G)$ and $\Delta p_{i\epsilon} = (\Delta p_i)_\epsilon$ gives $\|\nabla p_i - \nabla p_{i\epsilon}\|_{F^q} \to 0$.

iii) Let now $G$ be an exterior domain and for $k \in \mathbb{N}$ let $R_k := \{x \in \mathbb{R}^n : k < |x| < 2k\}$. Then the Poincaré-inequality holds for $v \in H^{1,q}(R_1)$ with a certain constant $C = C(q,n) > 0$.
Then with the same constant we get by means of a scaling argument

$$\|v\|_{q,R_k} \le k \cdot C \|\nabla v\|_{q,R_k} \quad \text{for all } v \in H^{1,q}(R_k) \text{ with } \int_{R_k} v\,dx = 0.$$

Let $\rho \in C_o^\infty(\mathbb{R}^n)$, $0 \le \rho \le 1$, $\rho(x) = 1$ for $|x| \le 1$ and $\rho(x) = 0$ for $|x| \ge 2$. Put $\rho_k(x) := \rho(k^{-1}x)$. Given $\nabla p \in F^q(G)$, put $c_k := |R_k|^{-1} \int_{R_k} p\,dx$ and $p_k(x) := \rho_k(x)(p(x)-c_k)$. Then $\nabla p_k = (\nabla\rho_k)(p-c_k) + \rho_k \cdot \nabla p$. If $\|\nabla\rho\|_\infty \le M$, then since $\text{supp}(\nabla\rho_k) \subset R_k$ we get by the inequality above

$\|(\nabla\rho_k)(p-c_k)\|_{q,G} \le k^{-1}M\|p-c_k\|_{q,R_k} \le CM\|\nabla p\|_{q,R_k} \to 0$ $(k \to \infty)$. Since $\rho_k \cdot \nabla p \to \nabla p$ in $L^q(G)$ we see $\|\nabla p-\nabla p_k\|_{q,G} \to 0$. Since $\Delta p_k = \Delta\rho_k(p-c_k) + 2\nabla\rho_k\nabla p + \rho_k\Delta p$ we readily see by the inequality above $\|\Delta\rho_k(p-c_k)\|_q \longrightarrow 0$ and therefore $\|\Delta p-\Delta p_k\|_q \longrightarrow 0$. For fixed $j \in \mathbb{N}$ consider $G_j := G \cap \{x\colon |x| \le j\}$. Then $p_k \in F^q(G_{4k})$ and vanishes in $G_{4k}\backslash G_{2k}$. Because of this property applying the approximation procedure from part ii) to $p_k$ in $G_{4k}$ we find a sequence $p_{ki} \in C_o^\infty(\bar{G}_{4k})$ with $\|\nabla p_{ki}-\nabla p_k\|_{F^q(G_{3k})} \longrightarrow 0$ such that $p_{ki}$ vanishes in $G_{4k}\backslash G_{3k}$. Extending $p_{ki}$ by zero to $G$ we are finished. □

We need some well known facts about traces from functions in Sobolev spaces. We use the Sobolev spaces $W^{1-1/s,s}(\partial G)$ with norm $\|.\|_{1-1/s,s},\partial G$ (compare e.g. [5] p. 337 and p. 341 or [11] p. 99 and p. 103). Let $G$ be a bounded domain with boundary $\partial G \in C^1$. Then for $1 < s < \infty$ there exists a map $V(=V_s) : H^{1,s}(G) \longrightarrow W^{1-1/s,s}(\partial G)$ such that $V\,v = v|_{\partial G}$ if $v \in C^1(\bar{G})$ and with a constant $K_s = K(s,\partial G)$ we have

(4.1) $\|Vv\|_{1-1/s,s,\partial G} \le K_s\|v\|_{1,s,G}$ for $v \in H^{1,s}(G)$.

$Vv$ is called the trace of $v$ on $\partial G$. If $v \in H_o^{1,s}(G)$ then $Vv = 0$. Further there exists an extension operator $T(=T_s) : W^{1-1/s,s}(\partial G) \longrightarrow H^{1,s}(G)$ with the following properties: $VTw = w$ for $w \in W^{1-1/s,s}(\partial G)$, $Tw \in C^1(\bar{G})$ if $w \in C^1(\partial G)$, $(TVv-v) \in H_o^{1,s}(G)$ for $v \in H^{1,s}(G)$ and with a constant $K'_s = K'_s(s,\partial G)$ we have

(4.2) $\|Tw\|_{1,s,G} \le K'_s\|w\|_{1-1/s,s,\partial G}$ for $w \in W^{1-1/s,s}(\partial G)$.

Since $C^\infty(\bar{G})$ is dense in $H^{1,s}(G)$ given $w \in W^{1-1/s,s}(\partial G)$, there exists a sequence $(v_i)$ in $C^\infty(\bar{G})$ such that $\|Tw-v_i\|_{1,s,G} \longrightarrow 0$. Since $V(Tw-v_i) = w-v_i|_{\partial G}$ by (4.1) $v_i|_{\partial G} \longrightarrow w$ in $W^{1-1/s,s}(\partial G)$. Therefore $\{v|_{\partial G} : v \in C^1(\bar{G})\}$ is dense in $W^{1-1/s,s}(\partial G)$.

By $W^{-1/s',s'}(\partial G) := (W^{1-1/s,s}(\partial G))^*$ we denote the dual space of $W^{1-1/s,s}(\partial G)$ $(s' = \frac{s}{s-1})$ equipped with norm $(g \in W^{1-1/s,s'}(\partial G))$

$$\|g\|_{-1/s',s',\partial G} := \sup\left\{\frac{|\langle g,w\rangle|}{\|w\|_{1-1/s,s,\partial G}} : 0 \ne w \in W^{1-1/s,s}(\partial G)\right\}$$

We sketch now the necessary technical changes for exterior domains.
If $G$ is an exterior domain, $G = \mathbb{R}^n \setminus K$ where $K \neq \emptyset$ is compact, we choose $R > 0$ such that $K \subset B_R$ where $B_R := B_R(0)$. Let $G_0 := G \cap B_{2R}$ and choose $\rho \in C_0^\infty(B_{2R})$ such that $\rho = 1$ on $B_R$. The considerations above we apply to $G_0$ and define now $Vv := V_{G_0}(\rho v)$ where $V_{G_0}$ denotes the trace operator with respect to $G_0$. Since $\rho = 0$ on $\partial B_{2R}$, $V_{G_0}(\rho v)|_{\partial B_{2R}} = 0$. Since $\|\rho v\|_{1,s,G} \le C(R)\|v\|_{1,s,G_0}$ we get from (4.1) with $\tilde{K} := K\,C(R)$

(4.1′) $\|Vv\|_{1-1/s,\partial G} \le \tilde{K}_s \|v\|_{1,s,G_0}$ for $v \in H^{1,s}(G_0)$.

Conversely given $w \in W^{1-1/s}(\partial G)$ we extend it by zero to $\partial B_{2R}$ and define $Tw := \rho T_{G_0} w$. Then $\|Tw\|_{1,s,G} = \|\rho T_{G_0} w\|_{1,s,G_0} \le C(R)\|T_{G_0} w\|_{1,s,G_0}$ and from (4.2) follows with $\tilde{K}'_s := K'_s \cdot C(R)^{-1}$

(4.2′) $\|Tw\|_{1,s,G} \le \tilde{K}'_s \|w\|_{1-1/s,s,\partial G}$.

If $w \in C^1(\partial G)$ then $T_{G_0} w \in C^1(\bar{G}_0)$ and therefore $Tw = \rho T_{G_0} w \in C_0^1(\bar{G})$. We want to show now $(TVv - v) \in H_0^{1,s}(G)$ for $v \in H^{1,s}(G)$. Let $0 < R' < R$ be such that $K \subset B_{R'}$ and choose $\sigma \in C_0^\infty(B_R)$ such that $\sigma = 1$ in $B_{R'}$. Given $v \in H^{1,s}(G)$ then $\sigma v \in H^{1,s}(G)$ and $T_{G_0} V(\sigma v) = T_{G_0} V_{G_0}(\rho\sigma v) = T_{G_0} V_{G_0}(\sigma v)$ since $\rho\sigma = \sigma$.
Therefore $T_{G_0} V(\sigma v) - \sigma v \in H_0^{1,s}(G_0)$ we have
$TV(\sigma v) - \sigma v = \rho(T_{G_0} V(\sigma v) - \sigma v) \in H_0^{1,s}(G_0) \subset H_0^{1,s}(G)$. The function $(1-\sigma)v$ vanishes in a neighborhood of $\partial G$ and may be extended by zero to an element of $H^{1,s}(\mathbb{R}^n)$. Since $H^{1,s}(\mathbb{R}^n) = H_0^{1,s}(\mathbb{R}^n)$ we get $(1-\sigma)\,v \in H_0^{1,s}(G)$. Then $V((1-\sigma)v) = 0$ and therefore $TVv - v = (TV(\sigma v) - \sigma v) - (1-\sigma)v$ belongs to $H_0^{1,s}(G)$.
For $T_{G_0} w$ there exists $\tilde{v}_i \in C^\infty(\bar{G}_0)$ such that $\|T_{G_0} w - \tilde{v}_i\|_{1,s,G_0} \to 0$. Put $v_i := \rho\tilde{v}_i$ then $\|Tw - v_i\|_{1,s,G} \to 0$ and $v_i$ belongs to $C_0^\infty(\bar{G})$. Therefore $\{v|_{\partial G} : v \in C_0^1(\bar{G})\}$ is dense in $W^{1-1/s,s}(\partial G)$ if $G$ is an exterior domain.
If $G$ is bounded and $\nabla p \in E^s(G)$ then $p \in H^{1,s}(G)$. If $G$ is an exterior domain and $\nabla p \in E^s(G)$ then $p|_{G_0} \in H^{1,s}(G_0)$. Therefore in both cases $Vp \in W^{1-1/s,s}(\partial G)$. Conversely for $w \in W^{1-1/s,s}(\partial G)$ by construction of $T$ we have $\nabla Tw \in E^s(G)$.

We extend now the concept of trace at the boundary to that of the normal derivative for elements of $F^q(G)$. In the more general situation of those vector fields $u \in L^q(G)^n$ such that div $u \in L^q(G)$, for G bounded this was done for $q = 2$ by Temam [17], p. 9 - 13 and for general $1 < q < \infty$ by Fujiwara-Morimoto [2], p. 686 - 692. Our procedure is slightly different. For bounded as well as for exterior domains and for $1 < q < \infty$ we first construct the trace operator for the normal derivative at the boundary for elements of $F^q(G)$ and later (Theorem 5.3) we extend this concept to vector fields such as mentioned above by means of the Helmholtz decomposition.

<u>Theorem 4.2.</u> Let $G \subset \mathbb{R}^n$ be either a bounded or an exterior domain with $\partial G \in C^1$ and let $1 < q < \infty$. Then there exists a map $S_q : F^q(G) \longrightarrow W^{-1/q,q}(\partial G)$ such that $S_q(\nabla p) = \partial_N p|_{\partial G} \equiv \langle N, \nabla p|_{\partial G}\rangle$ for $\nabla p \in E^\infty(G)$. We write $\partial_N p|_{\partial G} := S_q(\nabla p)$ for

$\nabla p \in F^q(G)$ too. Further with $\tilde{K}_{q'} > 0$ by (4.1′), for $\nabla p \in F^q(G)$

$$(4.3) \qquad \|\partial_N p|_{\partial G}\|_{-1/q,q,\partial G} \le \tilde{K}_{q'} \|\nabla p\|_{F^q} = \tilde{K}_{q'} (\|\nabla p\|_q^q + \|\Delta p\|_q^q)^{1/q}$$

For $w \in H^{1,q'}(G)$ let $w|_{\partial G}$ be the trace of w. Then for $\nabla p \in F^q(G)$ and $w \in H^{1,q'}(G)$ the generalized Gauß-Green formula holds:

$$(4.4) \qquad \langle \partial_N p|_{\partial G}, w|_{\partial G}\rangle_{\partial G} = \langle \nabla p, \nabla w\rangle_G + \langle \Delta p, w\rangle_G$$

<u>Proof.</u> Let $\nabla p \in E^\infty(G)$ and $w \in C^1(\partial G)$. Then $Tw \in C^1(\bar{G})$ (or $Tw \in C_0^1(\bar{G})$ if G is an exterior domain). By the classical Gauß-Green formula (4.4) holds for Tw and p and we get

$$(4.5) \qquad |\langle \partial_N p|_{\partial G}, w|\partial G\rangle_{\partial G}| \le \|\nabla p\|_{q,G}\|\nabla Tw\|_{q',G} + \|\Delta p\|_{q,G}\|Tw\|_{q',G}$$
$$\le \|\nabla p\|_{F^q} \|Tw\|_{1,q',G} \le$$
$$\le \tilde{K}_{q'}\|\nabla p\|_{F^q} \|w\|_{1-1/q',q',\partial G}$$

Therefore by $\langle g,w\rangle := \langle \partial_N p|_{\partial G}, w|_{\partial G}\rangle_{\partial G}$ a linear functional continuous with respect to $W^{1-1/q',q'}(\partial G)$-norm on the dense subspace $C^1(\partial G)$ is defined. There exists a unique normpreserving extension $\tilde{g}$ on the whole space $W^{1-1/q',q'}(\partial G)$ and we define $S_q(\nabla p) := \tilde{g}$. Since $E^\infty(G)$ is dense in $F^q(G)$ and $\|S_q(\nabla p)\|_{-1/q,q,\partial G} \le \tilde{K}_{q'}\|\nabla p\|_{F^q}$ by (4.5) there exists a unique extension of the operator $S_q$ to $F^q(G)$ defined by $S_q(\nabla p) := \lim_{i\to\infty} S_q(\nabla p_i)$ if $\|\nabla p - \nabla p_i\|_{F^q} \longrightarrow 0$.

Clearly (4.3) holds. Given $\nabla p \in F^q(G)$ and $w \in H^{1,q'}(G)$, put

$v := w\big|_{\partial G} \in W^{1-1/q',q'}(\partial G)$. Then there are $(\nabla p_i) \subset F^\infty(G)$ and $v_i \in C^1(\partial G)$ such

that $\|\nabla p_i - \nabla p\|_{F^q} \to 0$ and $\|v_i - v\|_{1-1/q',q'\partial G} \to 0$. Since $Tv_i \in C^1(\bar{G})$

(respectively $Tv_i \in C_0^1(\bar{G})$ if G exterior) we get

$$(4.6) \qquad \langle \partial_N p\big|_{\partial G}, \ v\big|_{\partial G}\rangle_{\partial G} = \lim_{i\to\infty}\langle \partial_N p_i\big|_{\partial G}, \ v_i\big|_{\partial G}\rangle_{\partial G} =$$
$$= \lim_{i\to\infty}(\langle \nabla p_i, \nabla Tv_i\rangle_G + \langle \Delta p_i, Tv_i\rangle_G) =$$
$$= \langle \nabla p, \nabla Tv\rangle_G + \langle \Delta p, Tv\rangle_G$$

Since $Vw = v$ we have $z := Tv - w = TVw - w \in H_0^{1,q'}(G)$. Then there is a sequence

$(z_i) \subset C_0^\infty(G)$ such that $\|z - z_i\|_{1,q',G} \to 0$ and therefore $\langle \nabla p, \nabla z\rangle_G + \langle \Delta p, z\rangle_G =$

$= \lim_{i\to\infty}(\langle \nabla p, \nabla z_i\rangle_G + \langle \Delta p, z_i\rangle_G) = 0$ and (4.4) follows from (4.6). $\square$

Next corollary gives a uniqueness result for the weak Neumann problem in $L^q$.

**Corollary 4.3.** Let $G \subset R^n$ be either a bounded or an exterior domain with
boundary $\partial G \in C^1$. Let $1 < q < \infty$ and $\nabla p \in F^q(G)$ with $\Delta p = 0$ and $\partial_N p\big|_{\partial G} = 0$.
Then $\nabla p = 0$.

**Proof.** By (4.4) we get $\langle \nabla p, \nabla w\rangle_G = 0$ especially for all $\nabla w \in E^\infty(G)$. Since
$E^\infty(G)$ is dense in $E^{q'}(G)$ with respect to $\|\nabla\|_{q'}$-norm we conclude $\langle \nabla p, \nabla \psi\rangle = 0$
for all $\nabla \psi \in E^{q'}(G)$ and by Lemma 3.9 follows $\nabla p = 0$. $\square$

A first application is the weak solution of the Neumann problem.

**Theorem 4.4.** Let $G \subset R^n$ be a bounded or an exterior domain with boundary
$\partial G \in C^1$. Let $1 < q < \infty$ and let $f \in L^q(G)$, $g \in W^{-1/q,q}(\partial G)$ be given. If G is
unbounded assume in addition that there is $R > 0$ such that $f = 0$ a.e. on
$G \cap \{x : |x| \geq R\}$. Put $G_R := G \cap \{x : |x| < R\}$ and $G_R := G$ if G is bounded.
Further assume

$$(4.7) \qquad \langle g, 1\rangle_{\partial G} - \int_{G_R} f \, dx = 0, \text{ where 1 denotes the function } \phi \equiv 1.$$

Then there exists a unique $\nabla p \in L^q(G)$ with $p \in L_{loc}^q(\bar{G})$ such that

$$(4.8) \qquad \langle \nabla p, \nabla \phi\rangle = \langle g, \phi\big|_{\partial G}\rangle - \langle f, \phi\rangle_{G_R}$$

for all $\phi \in C_0^1(\bar{G})$. Further $\Delta p = f$ in the sense of distributions and $\partial_N p\big|_{\partial G} = g$

In addition there is a constant $\gamma_1 = \gamma_1(G,R,q) > 0$ such that with $C > 0$ by Theorem 1.3

(4.9) $$\|\nabla p\|_q \le C\,\gamma_1(\|g\|_{-1/q,q,\partial G} + \|f\|_{q,G}).$$

<u>Proof.</u> If G is an exterior domain, $G = \mathbb{R}^n\setminus K$, with K compact, assume without loss of generality that $K \subset B_R$ is satisfied. Then $G_R$ is a domain with $\partial G_R \in C^1$ and the Poincaré inequality holds with a constant $d > 0$. If $\nabla\phi \in E^{q'}$ then $\phi|_{G_R} \in H^{1,q'}(G_R)$ and therefore

$$F^*(\phi) := \langle g,\phi|_{\partial G}\rangle - \langle f,\phi\rangle_{G_R}$$

is well defined for all $\phi$ with $\nabla\phi \in E^{q'}(G)$.

If $\int_{G_R}\phi dx = 0$ then by (4.1′) and the Poincaré inequality

$$|\langle g,\phi|_{\partial G}\rangle| \quad\le\quad \tilde K_{q'}\|g\|_{-1/q,q,\partial G}\,\|\phi\|_{1,q',G_R} \quad\le\quad \|g\|_{-1/q,q,\partial G}\,\tilde K_{q'}(1+d^{q'})^{\frac{1}{q'}}\|\nabla\phi\|_{q',G}$$

and

$|\langle f,\phi\rangle_{G_R}| \le \|f\|_{q,G_R}\|\phi\|_{q',G_R} \le d\|f\|_{q,G_R}\|\nabla\phi\|_{q',G}$. Put $\gamma := \max(d,\tilde K_{q'}(1+d^{q'})^{\frac{1}{q'}})$.

If $\phi$ with $\nabla\phi \in E^{q'}(G)$ is arbitrary then put $\tilde\phi := \phi - c(\phi)$ where $c(\phi) := |G_R|^{-1}\int_{G_R}\phi\,dx$. By (4.7) $F^*(c(\phi)) = 0$ and we end with

$|F^*(\phi)| = |F^*(\phi - c(\phi))| \le \gamma(\|g\|_{-1/q,\partial G} + \|f\|_{q,G_R})\ \|\nabla\phi\|_{q',G}$. Therefore by $F^*$ a continuous linear functional on $E^{q'}(G)$ is defined and by Theorem 1.3 a unique $\nabla p \in E^q(G)$ exists such that (4.8) and (4.9) hold. If $\phi \in C_o^\infty(G)$ then $\langle g,\phi|_{\partial G}\rangle = 0$ and $\langle\nabla p,\nabla\phi\rangle = -\langle f,\phi\rangle$ which proves $\Delta p = f$ (in the sense of distributions) and $\nabla p \in F^q(G)$. Then by (4.4) and (4.8) $\langle\partial_N p|_{\partial G},\ \phi|_{\partial G}\rangle = \langle g,\phi|_{\partial G}\rangle$ for all $\phi$ with $\nabla\phi \in E^{q'}(G)$. Since $W^{1-1/q',q'}(\partial G) = \{\phi|_{\partial G} : \nabla\phi \in E^{q'}(G)\}$ we find $\partial_N p|_{\partial G} = g$. □

<u>Remark 4.5.</u> For bounded G condition (4.7) is necessary too. It is the generalized version of the classical necessary condition for the Neumann problem "$\Delta p = f$ in G, $\partial_N p = g$ at $\partial G$" to be solvable. Clearly if $\nabla p \in E^q(G)$ and (4.8) holds, then for $\phi \equiv 1$ (4.7) follows.

<u>Corollary 4.6.</u> *Let $G \subset \mathbb{R}^n$ be a bounded or an exterior domain with $\partial G \in C^1$. Then for each $g \in W^{-1/q,q}(\partial G)$ there is $\nabla p \in F^q(G)$ such that $\partial_N p|_{\partial G} = g$ (that is the map $S_q : F^q(G) \to W^{1-1/q,q}(\partial G)$ defined by Theorem 4.2 is surjective).*

$\nabla p \in F^q(G)$ may be chosen such that $\Delta p = \langle g,1\rangle |G_R|^{-1}\chi_{G_R}$ (where $\chi_{G_R}$ denotes the characteristic function of $G_R$) and therefore with a constant $\gamma_2 > 0$

(4.10) $\|\nabla p\|_{F^q} \le \gamma_2 \|g\|_{-1/q,q,\partial G}$

Further $\{\partial_N p|_{\partial G} : \nabla p \in E^\infty(G)\}$ is dense in $W^{-1/q,q}(\partial G)$. That means $C^1(\partial G)$ is dense in $W^{-1/q,q}(\partial G)$.

Proof. Choose $f := \langle g,1\rangle |G_R|^{-1} \chi_{G_R}$. Then $f \in L^q(G_R)$ and (4.7) holds.

Since $|\langle g,1\rangle| \le \tilde{K}_{q'} |G_R|^{1/q'} \|g\|_{-1/q,q,\partial G}$ we see $\|f\|_q \le \tilde{K}_q \cdot \|g\|_{-1/q,q,\partial G}$

and (4.10) follows from (4.9). Since $E^\infty(G)$ is dense in $F^q(G)$ the density property is clear. □

## 5. Proof of the Helmholtz decomposition (Theorem 1.4) and normal components of vector fields at the boundary.

The following proof of Theorem 1.4 is solely based on Theorem 1.1, Corollary 1.2 and Theorem 1.3.

Proof of Theorem 1.4.

i) $u \in L^q_\sigma(G)$ implies $\langle u,\nabla\phi\rangle = 0$ for all $\nabla\phi \in E^{q'}(G)$:
Given $u \in L^q_\sigma(G)$ there is a sequence $(u_i)$ in $C^\infty_{0,\sigma}(G)$ such that $\|u-u_i\|_q \to 0$. For every $i \in \mathbb{N}$ $\langle u_i,\nabla\phi\rangle = -\langle \operatorname{div} u_i,\phi\rangle = 0$ for $\nabla\phi \in E^{q'}(G)$ and therefore $\langle u,\nabla\phi\rangle = \lim \langle u_i,\nabla\phi\rangle = 0$.

ii) $L^q_\sigma(G) \cap E^q(G) = \{0\}$: If $u$ belongs to this intersection then $u = \nabla p$ with $\nabla p \in E^q(G)$. Since $u \in L^q_\sigma(G)$ by i) $0 = \langle u,\nabla\phi\rangle = \langle\nabla p,\nabla\phi\rangle$ for all $\nabla\phi \in E^{q'}(G)$. By Theorem 1.3 $0 = \nabla p = u$.

iii) $L^q_\sigma(G) \oplus E^q(G)$ is closed: Clearly each of both spaces is closed (for $E^q(G)$ this follows from completeness of $E^q(G)$ with respect to the norm $\|\nabla .\|_q$).
Let $f_i = u_i + \nabla p_i$, $u_i \in L^q_\sigma(G)$, $\nabla p_i \in E^q(G)$ and assume $f_i \to f \in L^q(G)^n$.
By (1.3) and part i) of the proof

$$\|\nabla(p_i-p_k)\|_q \le C \sup_{0\ne\nabla\phi\in E^{q'}(G)} \langle \nabla(p_i-p_k), \nabla\phi\rangle \cdot \|\nabla\phi\|_{q'}^{-1}$$

$$= C \sup_{0\ne\nabla\phi\in E^{q'}(G)} \langle f_i-f_k,\nabla\phi\rangle \cdot \|\nabla\phi\|_{q'}^{-1} \le C \|f_i-f_k\|_q \to 0.$$

Since $E^q(G)$ is closed there is $\nabla p \in E^q(G)$ such that $\nabla p_i \longrightarrow \nabla p$. Clearly $u_i = f_i - \nabla p_i$ converges too and by closedness of $L^q_\sigma(G)$ there is $u \in L^q_\sigma(G)$ such that $u_i \longrightarrow u$. Then $f = u + \nabla p$.

iv) $L^q_\sigma(G) \oplus E^q(G) = L^q(G)^n$: Suppose $L^q_\sigma(G) \oplus E^q(G) \subsetneqq L^q(G)^n$.
Then by iii) and the Hahn-Banach theorem there is $F^* \in (L^q(G)^n)^*$ such that $F^*|_{L^q_\sigma(G)\oplus E^q(G)} = 0$ and $F^* \neq 0$. Since $(L^q(G)^n)^* \cong L^{q'}(G)^n$ isometrically isomorphic there is $u \in L^{q'}(G)^n$ such that $F^*(g) = \langle u,g\rangle$ for all $g \in L^q(G)^n$ and $\|u\|_{q'} = \|F^*\|^*_q > 0$. Since $F^*(\phi) = 0 = \langle u,\phi\rangle$ for all $\phi \in L^q_\sigma(G) \supset C^\infty_{0,\sigma}(G)$ by Theorem 1.1 and Corollary 1.2 there is $\nabla p \in E^{q'}(G)$ such that $\nabla p = u$. Since $\langle \nabla p,\nabla\psi\rangle = \langle u,\nabla\psi\rangle = F^*(\nabla\psi) = 0$ for all $\nabla\psi \in E^q(G)$ we conclude by Theorem 1.3 $\nabla p = 0$, that is $u = 0$ what contradicts $\|u\|_{q'} > 0$.

v) If $f = u + \nabla p$ then by i) $\langle f,\nabla\psi\rangle = \langle \nabla p,\nabla\psi\rangle$ for $\nabla\psi \in E^{q'}(G)$. If we put $P_q f := u$ then clearly $P_q^2 = P_q$.

vi) inequality (1.5): If $L^q(G)^n \ni f = u + \nabla p$, $u \in L^q_\sigma(G)$, $\nabla p \in E^q(G)$, then by Theorem 1.3

$$\|\nabla p\|_q \leq C \sup_{0\neq\nabla\psi\in E^{q'}(G)} \langle\nabla p,\nabla\psi\rangle \, \|\nabla\psi\|^{-1}_{q'} = C \sup_{0\neq\nabla\psi\in E^{q'}(G)} \langle f,\nabla\psi\rangle \, \|\nabla\psi\|^{-1}_{q'} \leq C \, \|f\|_q.$$

Therefore, $\|u\|_q = \|u + \nabla p - \nabla p\|_q \leq \|u + \nabla p\|_q + \|\nabla p\|_q \leq$

$$\leq \|f\|_q + C\,\|f\|_q = (C+1)\,\|f\|_q \quad \text{and} \quad \|u\|_q + \|\nabla p\|_q \leq (C+2)\,\|f\|_q.$$

vii) $L^q_\sigma(G)$-characterisation: Let $W^q(G) := \{u \in L^q(G) \mid \langle u,\nabla\psi\rangle = 0$ for all $\nabla\psi \in E^{q'}(G)\}$. Clearly by i) $L^q_\sigma(G) \subset W^q(G)$. If $u \in W^q(G)$, then by i) - iv) $u = v + \nabla p$, $v \in L^q_\sigma(G)$, $\nabla p \in E^q(G)$. For $\nabla\psi \in E^{q'}(G)$ we have again by i)

$$\langle\nabla p,\nabla\psi\rangle = \langle u,\nabla\psi\rangle - \langle v,\nabla\psi\rangle = 0.$$

Therefore from Theorem 1.3 $\nabla p = 0$, that is $u = v \in L^q_\sigma(G)$. □

<u>Remark 5.1.</u> Analyzing the proof above in fact we have proven more.
If $1 < q < \infty$ and G has property (3.1.s) for $s = q$ and $q'$ (compare section 3) then the Helmholtz decomposition holds true. As we have seen in section 3 e.g. the whole space, the half space and a "bended" half space have this property too. □

As in [2] we define for $1 < q < \infty$

(5.1) $\quad Y^q(G) := \{f \in L^q(G)^n : \operatorname{div} f \in L^q(G)\}.$

(5.2) $\quad \|f\|_{Y^q} := (\|f\|_q^q + \|\mathrm{div} f\|_q^q)^{\frac{1}{q}}.$

Clearly $Y^q(G)$ equipped with norm $\|.\|_{Y^q}$ is a (reflexive) Banach space.

Theorem 5.2. Let $G \subset \mathbb{R}^n$ be either a bounded or an exterior domain with boundary $\partial G \in C^1$ and let $1 < q < \infty$. Let $P_q : L^q(G)^n \to L^q_\sigma(G)$ be defined by Theorem 1.3.

i) If $u \in L^q_\sigma(G)$ then div $u = 0$, $u \in Y^q(G)$ and $\|u\|_q = \|u\|_{Y^q(G)}$. That is $P_q|_{Y^q(G)} : Y^q(G) \to L^q_\sigma(G)$ is onto.

ii) Let $f \in L^q(G)^n$. Then $f \in Y^q(G)$ if and only if $\nabla p := (I-P_q)f \in F^q(G)$ and $\Delta p = \mathrm{div}\, f$. That is $(I-P_q)\Big|_{Y^q(G)} : Y^q(G) \to F^q(G)$ is onto.

Furthermore we have the following direct ($q = 2$ : orthogonal) decomposition

$$Y^q(G) = L^q_\sigma(G) \oplus F^q(G)$$
$$f = u + \nabla p$$

where $u := P_q f$ and $\nabla p := (I-P_q)f$. In addition with $C > 0$ by Theorem 1.3 we get

(5.3) $\quad \|\nabla p\|_{F^q} = \|\nabla(I-P_q)f\|_{F^q} \le C\, \|f\|_{Y^q}$

(5.4) $\quad \|u\|_q = \|P_q f\|_q \le (C+1)\, \|f\|_{Y^q}$

(5.5) $\quad (\|u\|_q + \|\nabla p\|_{F^q}) \le (2C+1)\|f\|_{Y^q}$

iii) Let $Y^\infty(G) := C^\infty_{0,\sigma}(G) \oplus E^\infty(G)$. Then $Y^\infty(G)$ is dense in $Y^q(G)$ with respect to $\|.\|_{Y^q}$-norm.

Proof. If $u \in L^q_\sigma(G)$ then by (1.6) $\langle u,\nabla\phi\rangle = 0$ for all $\phi \in C^\infty_0(G)$ being equivalent to div $u = 0$ and therefore $u \in Y^q(G)$ and $\|u\|_q = \|u\|_{Y^q}$. If $\nabla p = (I-P_q)\, f \in E^q(G)$ then $\langle f,\nabla\phi\rangle = \langle \nabla p,\nabla\phi\rangle = -\langle p,\Delta\phi\rangle$ for $\phi \in C^\infty_0(G)$ and therefore $\Delta p = \mathrm{div}\, f \in L^q(G)$. Conversely if $f \in Y^q(G)$ we conclude $\nabla p \in F^q(G)$ and $\Delta p = \mathrm{div}\, f$. Clearly $F^q(G) \subset Y^q(G)$ and $(I-P_q)\nabla p = \nabla p$. The decomposition of $Y^q(G)$ follows immediately. (5.3) - (5.5) are an immediate consequence of (5.2) and (1.5). By Lemma 4.1 for $\nabla p := (I-P_q)\, f \in F^q(G)$ there is a sequence $(\nabla p_i) \subset E^\infty(G)$ such that $\|\nabla p-\nabla p_i\|_{F^q} \to 0$ and by definition for $u = P_q f \in L^q_\sigma(G)$ there exists a sequence $(u_i) \subset C^\infty_{0,\sigma}(G)$ such that $\|u-u_i\|_q \to 0$. Then $f_i := u_i + \nabla p_i \in Y^\infty(G)$ and $\|f-f_i\|_{Y^q} \to 0$. □

Clearly Theorem 5.2 implies that for bounded domains $C^\infty(\bar{G})^n$ is dense in $Y^q(G)$ and for exterior domains $C_0^\infty(\bar{G})^n$ is dense in $Y^q(G)$. For bounded domains this property was first proved in [2], p. 692. We extend now the concept of the trace of the normal derivative at the boundary for $\nabla p \in F^q(G)$ introduced in Theorem 4.2 to normal components of $f \in Y^q(G)$. This was originally done in [2], p. 686-692.

Theorem 5.3. Let $G \subset \mathbb{R}^n$ be either a bounded or an exterior domain with boundary $\partial G \in C^1$. Then there is a uniquely defined continuous linear operator $\tilde{S}_q : Y^q(G) \longrightarrow W^{-1/q,q}(\partial G)$ with the following properties.

i) $\tilde{S}_q f = f_N|_{\partial G}$ for $f \in Y^\infty(G)$ where $(f_N|_{\partial G})(x) := \langle N(x), f(x)|_{\partial G}\rangle$ denotes the classical normal component of f on $\partial G$ (we therefore write in the sequel $f_N|_{\partial G} := \tilde{S}_q f$ for $f \in Y^q(G)$).

ii) $\tilde{S}_q = S_q(I-P_q)$ where $S_q$ is defined by Theorem 4.2. $\tilde{S}_q$ maps $Y^q(G)$ onto $W^{-1/q,q}(\partial G)$.

iii) For $f \in Y^q(G)$, $f = u + \nabla p$ with $u \in L^q_\sigma(G)$ and $\nabla p \in F^q(G)$ we have $f_N|_{\partial G} = \partial_N p|_{\partial G}$ and

(5.6) $$\|f_N|_{\partial G}\|_{-1/q,q,\partial G} = \|\partial_N p|_{\partial G}\|_{-1/q,q,\partial G} \le \tilde{K}_{q'}\|\nabla p\|_{F^q} \le \tilde{K}_{q'}\cdot C\|f\|_{Y^q}$$

where $\tilde{K}_q$ is defined by Theorem 4.2 and C by Theorem 1.3.

iv) $L^q_\sigma(G) = \{u \in Y^q(G)\colon \operatorname{div} u = 0 \text{ and } u_N|_{\partial G} = 0\}$.

v) For $f \in Y^q(G)$ and for $w \in H^{1,q'}(G)$ the generalized Gauß-Green formula

(5.7) $$\langle f_N|_{\partial G}, w|_{\partial G}\rangle_{\partial G} = \langle f, \nabla w\rangle_G + \langle \operatorname{div} f, w\rangle_G$$

holds true.

Proof. By Lemma 5.2 ii) $(I-P_q) : Y^q(G) \longrightarrow F^q(G)$ is linear, continuous and onto. Then by Theorem 4.2 $\tilde{S}_q := S_q(I-P_q)$ maps $Y^q(G)$ linear and continuous onto $W^{-1/q,q}(\partial G)$. For $f \in Y^\infty(G)$, $f = u + \nabla p$ with $u \in C^\infty_{0,\sigma}(G)$ and $\nabla p \in F^\infty(G)$, we clearly have $f_N|_{\partial G} = \partial_N p|_{\partial G}$ in the classical sense. On the other hand

$S_q(I-P_q)f = \partial_N p|_{\partial G}$ and therefore $\tilde{S}_q f = f_N|_{\partial G}$ and $\tilde{S}_q$ has the desired property i). Since $Y^\infty(G)$ is dense in $Y^q(G)$ the operator $\tilde{S}_q$ is uniquely determined which proves ii). From definition of $f_N|_{\partial G}$, (4.3) and (5.3) we get (5.6) which proves iii). If $u \in L^q_\sigma(G)$ then by Theorem 5.2 i) $u \in Y^q(G)$, $\operatorname{div} u = 0$ and by definition $u_N|_{\partial G} = \tilde{S}_q u = 0$ since $(I-P_q)u = 0$. Conversely given $u \in Y^q(G)$ with $\operatorname{div} u = 0$ and $u_N|_{\partial G} = 0$ we write by Theorem 5.2 $u = v + \nabla p$, $v \in L^q_\sigma(G)$ and $\nabla p \in F^q(G)$, $\Delta p = \operatorname{div} u = 0$ and by part iii) of proof $\partial_N p|_{\partial G} = u_N|_{\partial G} = 0$ and therefore by Corollary 4.3 $\nabla p = 0$ that is $v \in L^q_\sigma(G)$. To prove (5.7) decompose $f = u + \nabla p$, $u \in L^q_\sigma(G)$, $\nabla p \in F^q(G)$ and observe $f_N|_{\partial G} = \partial_N p|_{\partial G}$, $\operatorname{div} f = \Delta p$ and $\langle u,\nabla w\rangle = 0$ by (1.6) since $\nabla w \in E^{q'}(G)$ for $w \in H^{1,q'}(G)$. Then (5.7) follows immediately from (4.4). □

An immediate consequence of Theorem 5.2 is the following most general version of the Neumann problem.

<u>Theorem 5.4.</u> Let $G \subset \mathbb{R}^n$ be either a bounded or an exterior domain with $\partial G \in C^1$. Let $1 < q < \infty$ and $f \in L^q(G)^n$ be given. Then there exists a unique (up to an additive constant) weak solution of the Neumann problem

$$\Delta p = \operatorname{div} f \text{ in } G,\ \partial_N p|_{\partial G} = f_N|_{\partial G}.$$

That is, there exists a unique $\nabla p \in E^q(G)$ such that

(5.8) $\quad \langle \nabla p,\nabla\phi\rangle = \langle f,\nabla\phi\rangle$ for all $\nabla\phi \in E^{q'}(G)$.

Futhermore, $(\nabla p-f) \in L^q_\sigma(G)$, $\operatorname{div}(\nabla p-f) = 0$ and $0 = (\nabla p-f)_N|_{\partial G} \in W^{-1/q,q}(\partial G)$.

<u>Proof.</u> Clearly the existence from $\nabla p \in E^q(G)$ with (5.8) follows from (1.3). Because of uniqueness of the Helmholtz decomposition, $u := (f-\nabla p) \in L^q_\sigma(G)$ and therefore $u \in Y^q(G)$, $\operatorname{div} u = 0$ and $u_N|_{\partial G} = 0$. □

Observe that in the theorem above neither $\partial_N p|_{\partial G}$ nor $f_N|_{\partial G}$ need to exist separately, solely $(\nabla p-f)_N|_{\partial G}$ exists. But if $f \in Y^q(G)$ then by Theorem 5.2 $\Delta p = \operatorname{div} f \in L^q(G)$ and $\partial_N p|_{\partial G} = f_N|_{\partial G} \in W^{-1/q,q}(\partial G)$. Concerning the Helmholtz decomposition we have therefore immediately the following

<u>Corollary 5.5.</u> Let the same assumptions as in Theorem 5.4. hold true. Then, given $f \in L^q(G)^n$, there exists a unique $u \in Y^q(G)$ with $\operatorname{div} u = 0$, $u_N|_{\partial G} = 0$ and a unique $\nabla p \in E^q(G)$ such that $f = u+\nabla p$ and $\operatorname{div}(f-\nabla p) = 0$ and $(f-\nabla p)_N|_{\partial G} = 0$

If moreover $f \in Y^q(G)$ then $\nabla p \in F^q(G)$, $\Delta p = \operatorname{div} f$ and $\partial_N p|_{\partial G} = f_N|_{\partial G}$.

Another easy consequence is the following density property which may be proven directly but with a considerable amount of work.

Theorem 5.6. Let G be a bounded or an exterior domain with $\partial G \in C^1$ and let $1 < q < \infty$. Then $\{\nabla p \in F^q(G) : \partial_N p|_{\partial G} = 0\}$ is dense in $E^q(G)$.

Proof. Since $E^\infty(G)$ is dense in $E^q(G)$ and $E^\infty(G) \subset F^q(G) \subset E^q(G)$ it suffices to prove density of $\{\nabla p \in E^\infty(G) : \partial_N p|_{\partial G} = 0\}$ in $F^q(G)$ with respect to $\|\nabla .\|_q$-norm. Choose a sequence $(\rho_k)$ in $C_0^\infty(G)$ with $0 \le \rho_k \le 1$ and $\rho_k \to 1$ in G. For $\nabla p \in F^q(G)$ put $f_k := \rho_k \cdot \nabla p$. Then $f_k \in Y^q(G)$ and $(f_k)_N|_{\partial G} = 0$ and $\nabla p_k := (I-P_k)\, f_k \in F^q(G), \partial_N p_k|_{\partial G} = 0$ and $\|\nabla p - \nabla p_k\|_q = \|(I-P_k)(\nabla p - f_k)\|_q \to 0$. □

## 6. The equivalence of the Helmholtz decomposition and Theorem 1.3.

We prove now that the Helmholtz decomposition and Theorem 1.3 are equivalent in the following sense:

Theorem 6.1. Let $G \subset \mathbb{R}^n$ be a domain such that $L^q(G)^n = L^q_\sigma(G) \oplus E^q(G)$ for a $q \in \mathbb{R}$ with $1 < q < \infty$ . With a constant $C > 0$ let inequalities (1.5) hold true.

i) Then for $q' = \frac{q}{q-1}$, the inequality

$$\|\nabla p\|_{q'} \le C \sup_{0 \ne \nabla\phi \in E^q(G)} |\langle \nabla p, \nabla\phi \rangle| \cdot \|\nabla\phi\|_q^{-1}$$

holds for all $\nabla p \in E^{q'}(G)$.

ii) For every continuous linear functional $F^*$ on $E^{q'}(G)$ there exists a uniquely determined $\nabla p \in E^q(G)$ such that $F^*(\nabla\phi) = \langle \nabla p, \nabla\phi \rangle$ for $\nabla\phi \in E^{q'}(G)$ and $\|\nabla p\|_q \le C\, \|F^*\|^*_{q'} \le C\, \|\nabla p\|_q$.

Proof. i) Let $\nabla p \in E^{q'}(G)$. Then

$$\|\nabla p\|_{q'} = \sup_{0 \ne f \in L(G)^n} {}_q \frac{|\langle \nabla p, f \rangle|}{\|f\|_q}.$$ Let $0 \ne f \in L^q(G)^n$ such that $\langle \nabla p, f \rangle \ne 0$,

$f = u + \nabla\phi$, $u \in L^q_\sigma(G)$, $\nabla\phi \in E^q(G)$ and $\|\nabla\phi\|_q \ne 0$. Then

$$\frac{|\langle \nabla p, f \rangle|}{\|f\|_q} = \frac{|\langle \nabla p, u \rangle + \langle \nabla p, \nabla\phi \rangle|}{\|f\|_q} = \frac{|\langle \nabla p, \nabla\phi \rangle|}{\|f\|_q} \le C\, \frac{|\langle \nabla p, \nabla\phi \rangle|}{\|\nabla\phi\|_q}$$

$$\leq C \sup_{0\neq\nabla\phi\in E^q(G)} |<\nabla p,\nabla\phi>|\cdot\|\nabla\phi\|_q^{-1}$$

and therefore $\|\nabla p\|_{q'} \leq C \sup_{0\neq\nabla\phi\in E^q(G)} |<\nabla p,\nabla\phi>|\cdot\|\nabla\phi\|_q^{-1}$.

ii) $E^{q'}(G)$ may be regarded as a closed linear subspace of $L^{q'}(G)^n$ and $F^*$ as a continuous linear functional on this subspace. By the Hahn-Banach theorem $F^*$ can be extended norm-preserving to a functional $\tilde{F}^*$ on $L^{q'}(G)^n$. Then there is a uniquely determined $f \in L^q(G)^n$ such that $\|f\|_q = \|\tilde{F}^*\|^*_{q'} = \|F^*\|^*_{q'}$ and $\tilde{F}^*(g) = <f,g>$ for all $g \in L^{q'}(G)^n$. For $\nabla\phi \in E^{q'}(G)$ we conclude $F^*(\nabla\phi) = <f,\nabla\phi>$.

Decompose $f = u+\nabla p$, $u \in L^q_\sigma(G)$, $\nabla p \in E^q(G)$. Since $<u,\nabla\phi> = 0$ we end with $F^*(\nabla\phi) = <\nabla p,\nabla\phi>$. By (1.5) $\|\nabla p\|_q \leq C\,\|f\|_q = C\,\|F^*\|^*_{q'}$. Clearly

$$\|F^*\|^*_{q'} = \sup_{0\neq\nabla\phi\in E^{q'}(G)} |F^*(\nabla\phi)|\,\|\nabla\phi\|_q^{-1} = \sup_{0\neq\nabla\phi\in E^{q'}(G)} |<\nabla p,\nabla\phi>|\,\|\nabla\phi\|_q^{-1} \leq \|\nabla p\|_q. \quad \square$$

Concerning the traces $\partial_N p|_{\partial G}$ we find

<u>Theorem 6.2.</u> Let the same assumptions as in Theorem 6.1 hold true. Assume in addition that there is an operator $\tilde{S}_q : Y^q(G) \to W^{-1/q,q}(\partial G)$ with the properties i) and iii) - v) from Theorem 5.3.

Then $S_q := \tilde{S}_q|_{F^q(G)}: F^q(G) \to W^{-1/q,q}(\partial G)$ has all the properties stated in Theorem 4.2.

<u>Proof:</u> If $\nabla p \in F^\infty(G) \subset Y^\infty(G)$ then by i) $\tilde{S}_q\nabla p = \partial_N p$ (the normal derivative). By iv) $\tilde{S}_q|_{L^q_\sigma(G)} = 0$ and therefore (since $Y^q(G) = L^q_\sigma(G) \oplus F^q(G)$)

$\tilde{S}_q|_{F^q(G)}: F^q(G) \to W^{-1/q,q}(\partial G)$ is onto.

The remaining properties are immediate. $\square$

<u>References:</u>


1. Adams, R.A.: Sobolev Spaces. New-York-San Francisco-London: Academic Press 1975.

2. Fujiwara, D. and Morimoto, H.: An $L_r$-theorem of the Helmholtz decomposition of vector fields. J. Fac. Sci. Univ. Tokyo, Sec.I 24, 685-700 (1977).

3. Gilbarg, D. and Trudinger, N.S.: Elliptic partial differential equations of second order. Berlin-Heidelberg-New York: Springer 1977.

4. Heywood, J.G.: On uniquenes questions in the theory of viscous flow. Acta Math. 136, 61-102 (1976).

5. Kufner, A., John,O. and Fučik, S.: Function spaces. Prague: Academia 1977.

6. Ladyzhenskaya, O.A.: The mathematical theory of viscous incompressible flow. New York-London-Paris: Gordon and Breach 1969.

7. Lions, J.L.: Quelques méthodes de résolution des problèmes aux limites non lineaires. Paris: Dunod 1969.

8. Lions, J.L. and Magenes, E.: Problemi ai limiti non omogenei (III). Ann. Scuola Norm. Sup. Pisa 15, 41-103 (1961).

9. Lions, J.L. and Magenes, E.: Problemi ai limiti non omogenei (V). Ibid. 16, 1-44 (1962).

10. Miyakawa, T.: On nonstationary solutions of the Navier-Stokes equations in an exterior domain. Hiroshima Math. J. 12, 115-140 (1982)

11. Nečas, J.: Les méthodes directes en théorie des équations elliptiques. Paris-Praque: Marsan, Academia 1967.

12. de Rham, G.: Variétés différentiables. Paris: Hermann 1960.

13. Schechter, M.: Coerciveness in $L^p$. Trans. Amer. Math. Soc. Vol. 197, 10-29 (1963).

14. Simader, C.G.: On Dirichlet's boundary value problem. Lecture Notes in Math. 268, Springer-Verlag 1972.

15. Simader, C.G. and Sohr, H.: The weak Dirichlet and Neumann problem in $L^p$ for the Laplacian in bounded and exterior domains. To appear.

16. Solonnikov, V.A.: Estimates for solutions of nonstationary Navier-Stokes equations. J. Soviet. Math. 8, 467-523 (1977).

17. Temam, R.: Navier-Stokes equations. Amsterdam-New York-Oxford: North Holland 1977.

18. Wahl von, W.: Abschätzungen für das Neumann-Problem und die Helmholtz-Zerlegung von $L^q$. To appear in: Nachr. Ak. d. Wiss. Göttingen, II. Mathematisch-Physikalische Klasse.

19. Yosida, K.: Functional Analysis. Grundlehren 123. Berlin-Heidelberg-New York: Springer 1965.

# On the Energy Equation and on the Uniqueness for *D*-Solutions to Steady Navier-Stokes Equations in Exterior Domains

Giovanni P. Galdi
Istituto di Ingegneria
Via Scandiana 21
Universitá degli Studi
Ferrara, Italia

**Introduction.**

Consider a viscous, homogeneous fluid $\mathcal{F}$ filling the region of the three-dimensional physical space $\mathbb{R}^3$ complementary to that occupied by one or more compact bodies $\mathcal{B}_i$, $i=1,\dots,N$. Denote by $\mathbf{v}=(v_1,v_2,v_3)$ and p the velocity and pressure field, respectively, in a given *time-independent* motion of $\mathcal{F}$. We shall assume throughout that $\mathbf{v}$ and p satisfy the following Navier-Stokes problem:

$$\left.\begin{array}{l}\Delta\mathbf{v} = \mathbf{v}\cdot\nabla\mathbf{v}+\nabla p-\mathbf{f}\\ \nabla\cdot\mathbf{v}=0\end{array}\right\}\text{ in }\Omega \tag{0.1}$$

with

$$\begin{array}{l}\mathbf{v}=\mathbf{v}_* \text{ at } \partial\Omega\\ \lim\limits_{|x|\to\infty}\mathbf{v}(x)=\mathbf{v}_\infty\end{array} \tag{0.2}$$

Here $\Omega\subset\mathbb{R}^3$ is the domain exterior to $\mathcal{B}_i$, $\mathbf{f}$ is the body force acting on the fluid and $\mathbf{v}_*$ and $\mathbf{v}_\infty$ are prescribed field (on $\partial\Omega$) and vector (in $\mathbb{R}^3$), respectively. Moreover,

$$\mathbf{v}\cdot\nabla\mathbf{v} \equiv \sum_{i=1}^{3} v_i\frac{\partial\mathbf{v}}{\partial x_i},\quad \nabla\cdot\mathbf{v}\equiv \sum_{i=1}^{3}\frac{\partial v_i}{\partial x_i}.$$

Without loss, we have set the coefficient of kinematic viscosity equal to one. A case of particular physical interest is when $\mathbf{v}_*\equiv 0$, $\mathbf{v}_\infty\neq 0$ and there is only one body $\mathcal{B}$ (say). In this situation the problem (0.1), (0.2) amounts to the determination, in a coordinate frame attached to $\mathcal{B}$, of the flow velocities in a steady motion of $\mathcal{B}$ through the fluid with velocity $-\mathbf{v}_\infty$. *In this (as well as in the following) article we shall always assume that* $\mathbf{v}_\infty\neq 0$ *and we take* $\mathbf{v}_\infty=\lambda\mathbf{e}_1$, $\lambda>0$, with $\mathbf{e}_1$ unit vector in the x-direction.

The fundamental contributions to the resolution of (0.1), (0.2) are mainly due to Leray (1933), Finn (1965) and Babenko (1973) who proved that, for *any* prescription of the data in a certain regularity class, problem (0.1), (0.2) admits at least one solution which behaves at infinity in such a way as to satisfy all requirements which are to be expected from the physical point of view[1]. In particular, this solution admits an asymptotic developement at infinity in which the dominant term is a solution to the corresponding *linearized* equations, namely, the *Oseen equations*. Therefore, it satisfies the *energy equation*[2]:

$$\int_{\Omega}\nabla v:\nabla v-\int_{\partial\Omega}[(v_*-v_\infty)\cdot T(v,p)-\tfrac{1}{2}(v_*-v_\infty)^2 v_*]\cdot n+\int_{\Omega}f\cdot(v-v_\infty)=0 \qquad (0.3)$$

with

$$T(v,p) = \{T_{ij}\} = \left\{p\delta_{ij}+\left(\frac{\partial v_i}{\partial x_j}+\frac{\partial v_j}{\partial x_i}\right)\right\}$$

*stress tensor*; furthermore, the solution exhibits a *wake region* $W$ in the direction of $v_\infty$, while the vorticity field decays exponentially fast outside $W$.

The basic steps in the proof of the above-said results can be sketched as follows. Without restrictions on the "size" of the data, Leray proved the existence of solutions to (0.1), (0.2) with velocity fields having a finite *Dirichlet integral*:

$$\int_{\Omega}\nabla v:\nabla v\le M \qquad (0.4)$$

where $M$ (>0) depends only on $f$, $v_*$ and $v_\infty$. For this reason Leray's solutions are also called *D-solutions*. The investigation of the asymptotic properties of *D*-solutions represented for quite a long time a formidable question to answer, to the point that, in 1959, R.Finn was led to introduce another class of solutions characterized by the requirement that the velocity field $v$ obeys, as $|x|\to\infty$, the condition

---

[1] If $v_\infty=0$, the same question presents several aspects which have not been clarified yet, see Galdi (1992b).

[2] As costumary, for $A=\{A_{ij}\}, B=\{B_{ij}\}$ second order tensors, we set $A:B=A_{ij}B_{ij}$, where summation over repeated indeces is understood. Moreover, the infinitesimal volume element in the integrals will be generally omitted.

$$v(x)-v_\infty = O(|x|^{-1/2-\varepsilon}), \quad \text{some } \varepsilon>0, \tag{0.5}$$

In a series of fundamental papers, Finn and his coworkers were then able to show that any such a solution possesses all the basic properties previously mentioned. For this reason, he called solutions satisfying (0.5) *physically reasonable* (*PR*). Moreover, in 1965, Finn showed that, if the data are "small enough" there exists a unique corresponding *PR*-solution and gave a detailed analysis of its asymptotic structure. The relation between a *D*-solution and a *PR*-solution was finally solved by K.I.Babenko (1973) who was able to demonstrate that every *D*-solution is a *PR*-solution, that is, if $v$ satisfies (0.4) then it necessarily satisfies (0.5) (the converse statement being easily shown).

It should be observed that, in spite of their relevance, the results just described present some undesired features. Actually, all "elementary" properties of *D*-solutions, such as their uniqueness or the validity of the energy balance (0.3), or even the simple Liouville-like theorem stating that every *D*-solution in the whole of $\mathbb{R}^3$ corresponding to $f\equiv 0$ is necessarily a constant, are obtained by employing their asymptotic properties as determined in the (rather involved) works of Babenko and Finn. This circumstance is certainly quite restrictive, since Babenko's result holds if the body force $f$ is of compact support. Therefore, we may wonder if there exists a different way of proving the above properties without resorting to the asymptotic structure of the solutions and, possibly, in a larger class of body forces.

The aim of this and of the following article by Galdi is to revisit the whole subject and to introduce a self-contained approach, completely different from that of the above authors, which avoids all the mentioned undesired features and which, in addition, furnishes a more direct proof of Babenko's result. Such an approach is based on a detailed $L^q$-theory of the Oseen equations which has been recently developed by Galdi (1992a). Specifically, in the present paper, we shall begin to derive some fundamental estimates for the nonlinear problem which will allow us, on the one hand, to establish the validity of the energy equation for any *D*-solution corresponding to data in a sufficiently large class and, on the other hand, to show its uniqueness in the class of *D*-solutions, provided the size of the data (in certain Lebesgue spaces $L^q$) is suitable restricted. Moreover, in the next article by Galdi we shall derive further estimates eventually leading to the proof that every *D*-solution is in fact a *PR*-solution.

The paper is organized as follows. In Section 1 we introduce notations and recall some well-known inequalities involving norms of functions in Sobolev spaces. In Section 2 we prove useful approximation theorems for solenoidal vector fields belonging to suitable homogeneous Sobolev spaces. In Section 3 we reproduce the results of Galdi (1992a) on the Oseen problem. In Section 4 we recall the definition of *D*-solution along with some related results. In Section 5 we give a generalized definition of energy equation which is appropriate for *D*-solutions and which, for sufficiently smooth data reduces to (0.3). We thus prove a quite general result ensuring that every *D*-solution, which enjoys some further summability conditions at large distances, necessarily obeys the generalized energy equation. On the basis of this result, we then show that if $\mathbf{f}\in L^{4/3}(\Omega)\cap L^{3/2}(\Omega)$ and $\mathbf{v}_*\in W^{4/3,3/2}(\partial\Omega)$[3], every corresponding *D*-solution satisfies the energy equation (0.3). Finally, in Section 6, we prove that if, in addition to the above assumptions on $\mathbf{f}$ and $\mathbf{v}_*$, it is also $\mathbf{f}\in L^{6/5}(\Omega)$ and the norms of $\mathbf{f}$ in $L^{6/5}(\Omega)$ and of $\mathbf{v}_*-\mathbf{v}_\infty$ in $W^{7/6,6/5}(\partial\Omega)$ are sufficiently small,then there is (one and) only one corresponding *D*-solution.

### 1. Notations and Preliminary Tools.

We begin by introducing some notations. $\mathbb{N}$ is the set of all positive integers. $\mathbb{R}^3$ is the three dimensional euclidean space and $\mathbb{R}$ is the real line. The surface of the unit ball centered at the origin is denoted by S.

By $\Omega$ we shall always denote a domain (open connected set) in $\mathbb{R}^3$. By $\bar{\Omega}$ we shall mean the closure of $\Omega$, by $\partial\Omega$ its boundary and by $\delta(\Omega)$ its diameter. We also set $\Omega^c=\mathbb{R}^3/\Omega$.

If $\Omega$ is a domain which is the complement of a (non-necessarily connected), compact set $\Omega^c$ with *bounded* boundary (*i.e.*, $\Omega$ is an *exterior* domain), taking the origin of coordinates into the interior of $\Omega^c$, for all

---

[3] This is the trace space of functions at the boundary, see Section 1.

$R>\delta(\Omega^c)$ and $R_2>R_1>\delta(\Omega^c)$ we set

$$\Omega_R=\{x\in\Omega:\ |x|<R\},\quad \Omega^R=\{x\in\Omega:\ |x|>R\},\quad \Omega_{R_1,R_2}=\{x\in\Omega:\ R_2>|x|>R_1\}.$$

We indicate by $C^k(\Omega)$ [resp. $C^k(\bar{\Omega})$], $0\le k\le\infty$, the set of functions in $\Omega$ which are continuous in $\Omega$ [resp. in $\bar{\Omega}$] with all their derivatives up to the order k (of arbitrary order for $k=\infty$). $C_0^k(\Omega)$ [resp. $C_0^k(\bar{\Omega})$] is the subclass of $C^k(\Omega)$ of functions of compact support in $\Omega$ [resp. the restriction to $\bar{\Omega}$ of functions from $C_0^k(\mathbb{R}^n)$].

For $\alpha=(\alpha_1,\alpha_2,\alpha_3)$, $\alpha_i\ge0$, let $D^\alpha=\partial^{\alpha_1+\alpha_2+\alpha_3}/\partial^{\alpha_1}x_1\partial^{\alpha_2}x_2\partial^{\alpha_3}x_3$. By $W^{m,q}(\Omega)$, $m\in\mathbb{N}\cup\{0\}$, $q\in[1,\infty]$, we denote the usual Sobolev space endowed with the norm

$$\|u\|_{m,q,\Omega}=\Big(\sum_{|\alpha|=0}^{m}\int_\Omega|D^\alpha u|^q\Big)^{1/q},$$

where the subscript $\Omega$ will be omitted if no confusion arises. We have $W^{0,q}(\Omega)=L^q(\Omega)$ and set $\|u\|_{0,q,\Omega}\equiv\|u\|_{q,\Omega}$. The duality pairing in $L^q(\Omega)$ is denoted by $(\cdot,\cdot)$, that is,

$$(u,v)=\int_\Omega uv.^4$$

The completion of $C_0^\infty(\Omega)$ in the norm of $W^{m,q}(\Omega)$ is denoted by $W_0^{m,q}(\Omega)$ and the dual space of $W_0^{m,q}(\Omega)$ is indicated by $W^{-m,q'}(\Omega)$, $q'=q/(q-1)$. Finally, by $W^{m-1/q,q}(\partial\Omega)$ we shall mean the trace space at $\partial\Omega$ of functions from $W^{m,q}(\Omega)$ and indicate by $\|\cdot\|_{m-1/q,q(\partial\Omega)}$ the associated norm, *cf.*, *e.g.*, Nečas (1967).

For $m\ge0$ and $1\le q<\infty$ we define the homogeneous Sobolev space

$$D^{m,q}=D^{m,q}(\Omega)=\{u\in L^1_{loc}(\Omega):D^\ell u\in L^q(\Omega),\ |\ell|=m\}.$$

One can prove that if $u\in D^{m,q}(\Omega)$ then

$$D^\ell u\in L^q(\Omega'),\ 0\le|\ell|\le m,\ \text{for all compact }\Omega'\text{ with }\bar{\Omega}'\subset\Omega,$$

or, in a shorter notation,

---

[4] We shall use the same notation for vectors **u**,**v** and second order tensors **A**,**B**, namely,

$$(u,v)=\sum_{i=1}^{3}\int_\Omega u_iv_i,\qquad (A,B)=\sum_{i,j=1}^{3}\int_\Omega A_{ij}B_{ij}.$$

Moreover, unless their use clarifies the context, the infinitesimal volume and surface elements in the integrals will be generally omitted.

$$u \in W^{m,q}_{loc}(\Omega).$$

If, in addition, $\Omega$ is locally lipschitzian, we also have $D^{\ell}u \in L^q(\Omega')$, $0 \le |\ell| \le m$, for all compact $\Omega' \subset \Omega$ or, in a shorter notation,

$$u \in W^{m,q}_{loc}(\bar{\Omega}).$$

Thus, if $\Omega$ is bounded and locally lipschitzian, $D^{m,q}(\Omega)$ is algebrically isomorphic to $W^{m,q}(\Omega)$.

In $D^{m,q}(\Omega)$ we introduce the seminorm

$$|u|_{m,q,\Omega} \equiv \Big( \sum_{|\ell|=m} \int_{\Omega} |D^{\ell}u|^q \Big)^{1/q} , \tag{1.1}$$

where, as before, the subscript $\Omega$ will be omitted if no confusion arises. *It is simple to show that* $\{D^{m,q}, |\cdot|_{m,q}\}$ *is a complete normed space,* provided we identify two functions $u_1, u_2 \in D^{m,q}(\Omega)$ whenever $|u_1-u_2|_{m,q}=0$, *i.e.*, $u_1$ and $u_2$ differ (at most) by a polynomial of degree m-1. Along with the spaces $D^{m,q}$ we shall consider the spaces $D^{m,q}_0 = D^{m,q}_0(\Omega)$ defined as the completion of $C^{\infty}_0(\Omega)$ in the seminorm (1.1). Clearly, one has

$$D^{m,q}_0(\Omega) \subseteq D^{m,q}(\Omega) \ , \ m \ge 0, \ q \ge 1,$$

and, further,

$$D^{0,q}_0(\Omega) = D^{0,q}(\Omega) = L^q(\Omega) \ , q \ge 1.$$

The dual space of $D^{1,q}_0(\Omega)$ is denoted by $D^{-1,q'}(\Omega)$ and the duality pairing between $D^{1,q}_0(\Omega)$ and $D^{-1,q}(\Omega)$ will be indicated by $[\cdot,\cdot]$.

We denote by $\mathcal{D}(\Omega)$ the set of solenoidal vector functions with components in $C^{\infty}_0(\Omega)$. Likewise, we let $\mathcal{D}^{1,q}_0(\Omega)$ be the completion of $\mathcal{D}(\Omega)$ in the norm $|\cdot|_{1,q}$. If $\Omega$ has a compact boundary and a mild degree of regularity, *e.g.*, $\Omega$ locally lipschitzian, one can prove the following characterization, *cf.* Bogovskiĭ (1980):

$$\mathcal{D}^{1,q}_0(\Omega) = \{u \in D^{1,q}_0(\Omega): \nabla \cdot u = 0\}, \ 1 \le q < \infty. \tag{1.1$'$}$$

$H_{0,q} = H_{0,q}(\Omega)$ denotes the completion of $C^{\infty}_0(\Omega)$ in the norm

$$\|u\|_q + \|\nabla \cdot u\|_q . \tag{1.2}$$

If $\Omega$ has a compact lipschitz boundary and $q \in (1,\infty)$, then the trace $\mathbf{u}\cdot\mathbf{n}$ of the normal component of $u \in H_{0,q}(\Omega)$ at $\partial\Omega$ is well defined as a member of the dual space of $W^{1-1/q,q}(\partial\Omega)$. Moreover, the following characterization holds, *cf.* *Temam (1977, Chapter I, §1.3)*,

$$H_{0,q}(\Omega) = \{u \in L^q(\Omega): \|\nabla \cdot u\|_q < \infty, \ \mathbf{u}\cdot\mathbf{n} = 0 \text{ at } \partial\Omega\}. \tag{1.2$'$}$$

We shall now collect some inequalities and properties of functions from $W^{1,q}$ and $D^{1,q}$ which will be frequently used in the sequel. We begin to recall the *Poincare's inequality*

$$\|u\|_{q,\Omega}\le c|u|_{1,q,\Omega}\,,\quad 1\le q<\infty, \tag{1.3}$$

holding for a bounded domain $\Omega$ and for all $u\in W_0^{1,q}(\Omega)$. The constant c varies with $\Omega$ and, in particular, the following estimate holds

$$c\le\beta|\Omega|^{1/3} \tag{1.3)'$$

where $\beta$ depends only on q, and $|\Omega|$ denotes the Lebesgue measure of $\Omega$. Another useful inequality which holds independently of the boundedness of $\Omega$ is the *Sobolev inequality*:

$$\|u\|_s\le[q/(3-q)\sqrt{3}]|u|_{1,q}\,,\quad 1\le q<3,\ s=3q/(3-q),\quad u\in D_0^{1,q}(\Omega). \tag{1.4}$$

The problem of extending the validity of (1.4) to the case when $\Omega$ is an exterior domain and u does not vanish at the boundary, has been considered by several authors. Here we shall reproduce a theorem which can be obtained by coupling a result of Galdi & Maremonti (1986) with one of Babenko (1973, Proposition 3), *cf.* Galdi (forthcoming, Theorem II.5.1).

**Theorem 1.1.** *Let $\Omega$ be an exterior locally lipschitzian domain. Assume* $u\in D^{1,2}(\Omega)$ *with*

$$\frac{\partial u}{\partial x_1}\in L^q(\Omega),\quad \textit{for some } q\in[1,2]$$

*Then, there exists* $\gamma_1>0$ *and* $u_0\in\mathbb{R}$ *such that*

$$w\equiv u-u_0\in L^r(\Omega),\quad r=3q,$$

*and the following inequality holds*

$$\|w\|_r^3\le\gamma_1\left\|\frac{\partial u}{\partial x_1}\right\|_q\left\|\frac{\partial u}{\partial x_2}\right\|_2\left\|\frac{\partial u}{\partial x_3}\right\|_2 \tag{1.5}$$

**Remark 1.1.** Inequality (1.5) is more general than inequality (1.4) with q=2.

We conclude this section with a result concerning approximations of functions of $D^{1,q}(\mathbb{R}^3)\cap L^r(\mathbb{R}^3)$ by smooth functions of compact support.

**Theorem 1.2** *Let* $u\in D^{1,q}(\mathbb{R}^3)\cap L^r(\mathbb{R}^3)$, $3\le q<\infty$, $1\le r_i<\infty$, $i=1,\dots k$. *Then, there exists a sequence* $\{u_m\}\subset C_0^\infty(\mathbb{R}^3)$ *such that*

$$v_m\to v \textit{ in } D^{1,q}(\mathbb{R}^3)\cap[\bigcap_{i=1}^k L^r(\mathbb{R}^3)].$$

*Proof.* The proof can be immediately achieved by using results due to

Sobolev (1963). Actually, in this paper it is proved that any function from $D^{1,q}(\mathbb{R}^3)$, $3\le q<\infty$, can be approximated by $C_0^\infty(\mathbb{R}^3)$ functions $u_{R,\varepsilon}$ of the type

$$u_{R,\varepsilon}=(\psi_R u)_\varepsilon \,, \tag{1.6}$$

where $\psi_R(x)$ is a suitable cut-off function which is 1 for $|x|<exp\sqrt{\ln R}$ and is zero for $|x|>R$ and $(\cdot)_\varepsilon$ denotes mollification in the sense of Friederichs. Clearly, the functions (1.6), for $R\to\infty$ and $\varepsilon\to0$, converge to u also in $L^{r_1}(\mathbb{R}^3)$, $1\le r_1<\infty$ and the theorem is completely proved.

## 2. Approximation Problems in Spaces $\mathcal{D}_0^{1,q}$.

A problem which we shall encounter in deriving energy inequality and uniqueness of $\mathcal{D}$-solutions is the following one. Assume

$$v\in\mathcal{D}_0^{1,q}(\Omega)\cap[\bigcap_{l=1}^{k} L^{r_l}(\Omega)], \quad 1<q,r_l<\infty.$$

Clearly, u can be approximated by a sequence $\{v'_m\}\subset\mathcal{D}(\Omega)$ (as a member of $\mathcal{D}_0^{1,q}(\Omega)$) and by a sequence $\{v''_m\}\subset C_0^\infty(\Omega)$ (as a member of $\bigcap_{l=1}^{k} L^{r_l}(\Omega)$). The question is now to ascertain if v can be approximated by the *same* sequence in *both* spaces or, in other words –taking into account that $\mathcal{D}(\Omega)\subset C_0^\infty(\Omega)$– if there is a sequence $\{v_m\}\subset\mathcal{D}(\Omega)$ such that as $k\to\infty$

$$v_m \longrightarrow v \text{ strongly in } \mathcal{D}_0^{1,q}(\Omega)\cap[\bigcap_{l=1}^{k} L^{r_l}(\Omega)]. \tag{2.1}$$

Of course this problem admits a trivial positive answer when $\Omega$ is bounded and q, $r_l$ are suitably related to each other. For example, take k=1 and assume that at least one of the following conditions is fulfilled:

*(i)* $r\le q$;

*(ii)* $q\ge3$;

*(iii)* $q<3$ and $r>3q/(3-q)$.

Then (2.1) follows at once. Actually, denoting by $\{v_m\}\subset\mathcal{D}(\Omega)$ a sequence converging to v in $\mathcal{D}_0^{1,q}(\Omega)$, in case *(i)* we have, by Hölder inequality and inequality (1.3),

$$\|v-v_m\|_r\le c|v-v_m|_{1,q} \longrightarrow 0.$$

In cases *(ii)* or *(iii)* the same conclusion can be drawn by using, instead of Hölder inequality, the Sobolev embedding theorem. Moreover, inequality

(1.4), gives the result also for *arbitrary* $\Omega$, provided $1<q<3$ and $r=3q/(3-q)$.

What can be said in the general case when q and $r_1$ are not necessarily related to each other? The aim of this section is to show that for $\Omega$ locally lipschitzian, it is always possible to find a sequence $\{v_m\}\subset\mathcal{D}(\Omega)$ satisfying (2.1)[5].

Our goal will be achieved through several intermediate steps. We begin to furnish suitable estimates for solutions to the equation $\nabla\cdot v=f$. We recall that a (bounded) domain $\Omega$ is said to be *star-shaped with respect to an open ball B* ($\subset\Omega$) if $\Omega$ is star-shaped with respect to every point of $B$.

We have

**Lemma 2.1.** *Let $\Omega$ be bounded and locally lipschitzian. Then there exists an open cover $\mathcal{G} = \{G_1,\dots,G_m,G_{m+1},\dots,G_{m+\nu}\}$ of $\bar{\Omega}$ such that*

(*i*) $\Omega_i\equiv\Omega\cap G_i$ *is star-shaped with respect to an open ball $B_i$ with $\bar{B}_i\subset\Omega_i$ for* $i=1,\dots,m+\nu$;

(*ii*) $\partial\Omega\subset\bigcup_{i=1}^{m}G_i$;

(*iii*) $G_i$ *is an open ball with $\bar{G}_i\subset\Omega$ for* $i=m+1,\dots,m+\nu$.

(*iv*) $\Omega = \bigcup_{i=1}^{m+\nu}\Omega_i$

*Moreover, if* $f\in C_0^\infty(\Omega)$ *with* $\int_\Omega f=0$, *we have* $f = \sum_{i=1}^{m+\nu} f_i$ *where*

(*v*) $f_i\in C_0^\infty(\Omega_i)$;

(*vi*) $\int_{\Omega_i} f_i=0$ ;

(*vii*) $f_i=\zeta f + \sum_{k=1}^{m_i}\theta_k\int_\Omega \phi_k f$, *where* $m_i\in\mathbb{N}$, $\zeta\in C_0^\infty(G_i)$, $\theta_k\in C^\infty(\Omega_i)$ *and* $\phi_k\in C_0^\infty(\bar{\Omega})$.

(*viii*) $\|f_i\|_{m,q}\le C\|f\|_{m,q}$, *for all* $m\ge0$ *and* $q\ge1$,

*where C depends only on* m, q *and* $\Omega$.

*Proof.* The result is essentially due to Bogovskiĭ (1980, Lemma 5). For the sake of completeness, we reproduce here a proof given by Galdi (forthcoming). We may find m locally lipschitzian domains $G_1,\dots,G_m$ such

[5] The results of the present section admit of a straightforward extension to the case of arbitrary space dimension $n\ge2$.

that $\Omega_i \equiv \Omega \cap G_i$ is star-shaped with respect to an open ball $B_i$ with $\overline{B}_i \subset \Omega_i$ and verifying, in addition, condition *(ii)*, *cf.*, *e.g*, Erig (1982, Lemma 5.3). Denote, next, by $G_0$ a domain with $\overline{G}_0 \subset \Omega$ such that $\{G_0, G_1, \ldots, G_m\}$ forms an open cover of $\overline{\Omega}$. Since $\Omega$ is bounded, we may find $\nu$ open balls $G_{m+1}, \ldots, G_{m+\nu}$ such that

$$\overline{G}_0 \subset \bigcup_{i=m+1}^{m+\nu} G_i$$

$$\left\{ \bigcup_{i=m+1}^{m+\nu} G_i \right\}^{-} \subset \Omega .$$

Evidently,

$$\mathcal{G} = \{G_1, \ldots, G_m, G_{m+1}, \ldots, G_{m+\nu}\}$$

is an open cover of $\Omega$ with $\overline{\Omega} \subset \bigcup_{i=1}^{m+\nu} G_i$. Setting

$$\Omega_i \equiv \Omega \cap G_i \,, \qquad i=1,\ldots,m+\nu$$

it is immediately obtained that properties *(i)*, *(iv)* are satisfied. Let us now pick $f \in C_0^\infty(\Omega)$ and show the existence of $f_i$ satisfying *(v)*–*(viii)*. To this end, let

$$\{\psi_1, \ldots \psi_{m+\nu}\}$$

be a partition of unity subordinate to $\mathcal{G}$, *cf.*,*e.g.*, Nečas (1967,Chapitre 1, Proposition 2.3), that is

(a) $\psi_i \in C_0^\infty(G_i)$, $i=1,\ldots,m+\nu$;

(b) $\sum_{i=1}^{m+\nu} \psi_i(x) = 1$, for all $x \in \Omega$.

Setting

$$D_2 = \bigcup_{i=2}^{m+\nu} \Omega_i \quad , \quad \Psi_2 = \sum_{i=2}^{m+\nu} \psi_i(x),$$

and observing that

$$\psi_1(x) + \Psi_2(x) = 1, \text{ for all } x \in \Omega,$$

we may write $f = f_1 + g_1$, where

$$f_1 = \psi_1 f - \chi_1 \int_\Omega \psi_1 f$$

$$g_1 = \Psi_2 f - \chi_1 \int_\Omega \Psi_2 f$$

and

$$\chi_1 \in C_0^\infty(\Omega_1 \cap D_2), \quad \int_\Omega \chi_1 = 1 \quad .$$

Since $\psi_1 \in C_0^\infty(G_1)$ and $f \in C_0^\infty(\Omega)$ it immediately follows that

$$f_1 \in C_0^\infty(\Omega_1).$$

Moreover, we have

$$\Psi_2 = \sum_{l=2}^{m} \psi_l(x) + \sum_{l=m+1}^{m+\nu} \psi_l(x)$$

so that, recalling the definition of $G_2,\dots,G_m,G_{m+1},\dots,G_{m+\nu}$ and that $f\in C_0^\infty(\Omega)$ it follows

$$g_1 \in C_0^\infty(D_2).$$

In addition, we have, evidently,

$$\int_{\Omega_1} f_1 = \int_{D_2} g_1 = 0.$$

We may then continue this procedure, by using $g_1$ as the function to be split. Precisely, setting

$$D_3 = \bigcup_{l=3}^{m+\nu} \Omega_l \quad , \quad \Psi_3 = \sum_{l=3}^{m+\nu} \psi_l(x),$$

we let

$$f_2 = \psi_2 g_1 - \chi_2 \int_\Omega \psi_2 g_1$$

$$g_2 = \Psi_3 g_1 - \chi_2 \int_\Omega \Psi_3 f$$

where

$$\chi_2 \in C_0^\infty(\Omega_2 \cap D_3)$$

$$\int_\Omega \chi_2 = 1 \quad .$$

From the expression of $g_1$ and property $(b)$ of the partition of unity we recover at once that f can be written as

$$f_2 = \psi_2(1-\psi_1)f - \chi_1\psi_2\int_\Omega (1-\psi_1)f - \chi_2\int_\Omega \psi_2(1-\psi_1)f + \chi_2\int_\Omega \psi_2\chi_1\int_\Omega (1-\psi_1)f,$$

which proves $f_2 \in C_0^\infty(\Omega_2)$, $(vii)$ and $(viii)$. Furthermore, it is

$$g_1 = f_2 + g_2 \, , \quad g_2 \in C_0^\infty(D_3),$$

$$\int_{\Omega_2} f_2 = \int_{D_3} g_2 = 0.$$

We may then use an iteration scheme to establish the validity of $(v)$–$(vii)$ for all $i=1,\dots,m+\nu$.

We have also

**Lemma 2.2.** *Let* $\Omega$, $G$ *be bounded, locally lipschitzian domains, with* $\Omega\cap G \equiv \Omega_0$ $(\neq\emptyset)$ *star-shaped with respect to a ball* $B$ *with* $\overline{B}\subset\Omega_0$. *Let, further,* $\phi_k$, $\theta_k$, $k=1,\dots,m$, $\zeta$ *and* $g$ *be functions such that*

$$\phi_k \in C_0^\infty(\overline{\Omega}),\ \theta_k \in C_0^\infty(\Omega_0),\ \zeta \in C_0^\infty(G),$$
$$g \in H_{0,q}(\Omega),\ 1<q<\infty, \quad \nabla\cdot g \in C_0^\infty(\Omega).$$

*Set*

$$f = \zeta\nabla\cdot g + \sum_{l=1}^{m} \theta_k \int_\Omega \phi_k \nabla\cdot g \tag{2.2}$$

*and suppose*

$$\int_{\Omega_0} f = 0. \tag{2.3}$$

*Then, there is at least one solution* **w** *to the problem*

$$\begin{aligned} &\nabla\cdot \mathbf{w} = f \\ &\mathbf{w} \in C_0^\infty(\Omega) \\ &\|\mathbf{w}\|_{1,s,\Omega_0} \le c\|\nabla\cdot g\|_{s,\Omega} \\ &\|\mathbf{w}\|_{q,\Omega_0} \le c\|g\|_{q,\Omega}\ , \end{aligned} \tag{2.4}$$

*where* s *is arbitrary in* $(1,\infty)$ *and* $c=c(\phi_k,\theta_k,\zeta,s,q,B,G,\Omega)$.

*Proof.* For simplicity, we shall restrict ourselves to discuss the case m=1 and set $\theta=\theta_1$, $\phi=\phi_1$. Clearly, $f\in C_0^\infty(\Omega_0)$. Then, by (2.2), (2.3) and by the assumption made on $\Omega_0$, we can find a solution to the problem by an explicit formula due to Bogovskiǐ (1979, eq.(6)), namely,

$$\mathbf{w}(x) = \int_{\Omega_0} f(y)\left[\frac{x-y}{|x-y|}\int_{|x-y|}^{\infty} \omega\!\left(y+\xi\frac{x-y}{|x-y|}\right)\xi^2 d\xi\right]dy \equiv \int_{\Omega_0} f(y)N(x,y)dy. \tag{2.5}$$

Here, $\omega$ is any function from $C_0^\infty(\mathbb{R}^n)$ such that

(*i*) $supp(\omega)\subset B$,

(*ii*) $\int_B \omega = 1$,

It is not difficult to show that the field (2.5) is in $C_0^\infty(\Omega_0)$, *cf.* Erig (1982, pp. 15, 16) and that it satisfies $(2.4)_1$, *cf.* Bogovskiǐ (1979, Lemma 1). Moreover, from Theorem 1 of Bogovskiǐ (1979) we also obtain for all $s\in(1,\infty)$

$$\|\mathbf{w}\|_{1,s} \le c_1\|f\|_s,$$

with $c_1=c_1(n,s,\Omega_0)$. Since

$$\|f\|_s \le c_2\|\nabla\cdot g\|_s,$$

with $c_2=c_2(\phi,\theta,\zeta,s,\Omega)$, also $(2.4)_3$ follows. It remains to show $(2.4)_4$. From (2.2) and (2.5) we find for i=1,2,3

$$w_l(x) = w_l^{(1)}(x) + w_l^{(2)}(x)$$

$$w_l^{(1)}(x) = \int_{\Omega_0} N_l(x,y)\zeta(y)D_j g_j(y)dy \tag{2.6}$$

$$w_l^{(2)}(x) = \left(\int_\Omega \phi\nabla\cdot g\right)\left(\int_{\Omega_0} N_l(x,y)\theta(y)dy\right)$$

Using the properties of $\omega$, we obtain

$$|N(x,y)| \le \left|\frac{x-y}{|x-y|}\int_{|x-y|}^{\infty} \omega\left(y+\xi\frac{x-y}{|x-y|}\right)\xi^{n-1}d\xi\right|$$

$$\le |x-y|^{1-n}\|\omega\|_\infty \int_0^{\delta(\Omega_0)} \xi^{n-1}d\xi \le c|x-y|^{1-n} \tag{2.7}$$

with $c=c(n,B,\Omega_0)$. Thus, from Young's inequality on convolutions it is

$$\left\|\int_{\Omega_0} N(x,y)\theta(y)dy\right\|_{q,\Omega_0} < \infty ,$$

and so

$$\|w^{(2)}\|_{q,\Omega_0} \le c_3\left|\int_\Omega \phi\nabla\cdot g\right|.$$

Since $g\in H_{0,q}(\Omega)$, the trace of the normal component of $g$ at $\partial\Omega$ is identically vanishing, *cf.* (1.2)′, and consequently we have

$$\int_\Omega \phi\nabla\cdot g = -\int_\Omega g\cdot\nabla\phi.$$

Therefore,

$$\|w^{(2)}\|_{q,\Omega_0} \le c_4\|g\|_{q,\Omega}. \tag{2.8}$$

Taking into account that $\zeta\in C_0^\infty(G)$, we may formally integrate by parts into $(2.6)_3$ to obtain

$$w_l^{(1)}(x) = -\lim_{\varepsilon\to 0}\left\{\int_{\Omega_0 - B_\varepsilon(x)} [N_l(x,y)g_j(y)D_j\zeta(y) + (D_jN_l(x,y))\zeta(y)g_j(y)]dy \right.$$
$$\left. + \int_{\partial B_\varepsilon(x)} N_l(x,y)\zeta(y)g_j(y)\frac{x_j-y_j}{|x-y|}\,d\sigma_y \right. \tag{2.9}$$
$$\equiv -\lim_{\varepsilon\to 0}\sum_{l=1}^{3} I_l(x,\varepsilon)$$

Since, by (2.7) and by Young's inequality on convolutions, it is

$$\left\|\int_{\Omega_0} N(x,y)g\cdot\nabla\zeta(y)dy\right\|_{q,\Omega_0} \le \|g\|_{q,\Omega} ,$$

we have, for a.a. $x\in\Omega$,

$$\lim_{\varepsilon\to 0} I_1(x,\varepsilon) = \int_{\Omega_0} N_l(x,y)g_j(y)D_j\zeta(y)dy . \tag{2.10}$$

Furthermore, by a simple computation we show for a.a. $x\in\Omega_0$

$$\lim_{\varepsilon\to 0} I_3(x,\varepsilon) = \zeta(x)g_j(x)\int_{\Omega_0} \frac{(x_i-y_i)(x_j-y_j)}{|x-y|^2}\,\omega(y)dy . \tag{2.11}$$

Finally, as we know from the work of Bogovskiĭ (1980), it is

$$D_jN_i(x,y) = K_{ij}(x,x-y) + G_{ij}(x,y)$$

where $K_{ij}$ is a Calderón-Zygmund kernel, while $|G_{ij}|$ is bounded by a weakly singular kernel. Therefore, from Calderón-Zygmund theorem we deduce for a.a. $x\in\Omega_0$

$$\lim_{\varepsilon\to 0} I_2(x,\varepsilon) = \int_{\Omega_0} K_{ij}(x,x-y)\zeta(y)g_j(y)dy + \int_{\Omega_0} G_{ij}(x,y)\zeta(y)g_j(y)dy, \tag{2.12}$$

where, of course, the first integral has to be understood in the Cauchy principal value sense. From (2.9)–(2.12), from Calderón-Zygmund theorem, and Young's inequality on convolutions we then conclude

$$\|w^{(1)}\|_{q,\Omega_0} \le c_6 \|g\|_{q,\Omega}. \tag{2.13}$$

Thus, estimates $(2.4)_4$ becomes a consequence of (2.6), (2.8) and (2.13) and the lemma is completely proved.

We are now in a position to show the following

**Theorem 2.1.** *Let $\Omega$ be bounded and locally lipschitzian domain. Then, given*

$$g\in H_{0,q}(\Omega),\ 1<q<\infty,$$

*there exists at least one solution* $v$ *to the problem*

$$\begin{aligned} &\nabla\cdot v = \nabla\cdot g \\ &v\in W_0^{1,q}(\Omega) \\ &\|v\|_{1,q} \le c\|\nabla\cdot g\|_q \\ &\|v\|_q \le c\|g\|_q \quad . \end{aligned} \tag{2.14}$$

*In particular, if* $\nabla\cdot g\in C_0^\infty(\Omega)$, *then* $v\in C_0^\infty(\Omega)$ *and we have*

$$\|v\|_{1,s} \le c\|\nabla\cdot g\|_s\ , \qquad \textit{for all } s\in(1,\infty). \tag{2.15}$$

*Proof*. We take first $g$ such that

$$g\in H_{0,q}(\Omega), \qquad \nabla\cdot g\in C_0^\infty(\Omega). \tag{2.16}$$

From Lemma 2.1, we can show the existence of N domains $\Omega_i=\Omega\cap G_i$, with $\{G_i\}$ an open covering of $\bar\Omega$, satisfying $(i)$–$(iv)$ of that lemma. Furthermore, we may write

$$\nabla\cdot g = \sum_{i=1}^{N} f_i$$

with f satisfying conditions (*v*)–(*viii*) of Lemma 2.1. From Lemma 2.2, for any i=1,...,N, we can determine the existence of a vector $v_i$ such that for all $s\in(1,\infty)$

$$\nabla\cdot v_i = f_i$$

$$v_i \in C_0^\infty(\Omega_i)$$

$$\|v_i\|_{1,s,\Omega_i} \le c\|\nabla\cdot g\|_{s,\Omega}$$

$$\|v_i\|_{q,\Omega_i} \le \|g\|_{q,\Omega}\ .$$

Thus, the field

$$v = \sum_{k=1}^{N} v_k$$

satisfies all requirements of the theorem which is thus proved if g satisfies (2.16). Assume, now, g merely belonging to $H_{0,q}(\Omega)$, $1<q<\infty$. By the definition of the space $H_{0,q}(\Omega)$, we can approximate g by a sequence $\{g_s\}\subset C_0^\infty(\Omega)$. For each s we then establish the existence of $v_s$ solving (2.14) with $g_s$ in place of g. Because of $(2.14)_{3,4}$ and of the weak compactness of the spaces $W_0^{1,q}(\Omega)$ and $L^q(\Omega)$, we attain the existence of $v\in W_0^{1,q}(\Omega)$ and of $\{v_{s'}\}\subset\{v_s\}$ such that

$$v_{s'} \to v \quad \text{weakly in } W_0^{1,q}(\Omega).$$

Evidently, v solves $(2.14)_1$ in the generalized sense and, again by $(2.14)_{3,4}$ (written for $v_{s'}$), v satisfies $(2.14)_{3,4}$. The proof of the theorem is therefore completed.

Our next objective is to show that every function from $D_0^{1,q}(\Omega)\cap L^r(\Omega)$, $q,r\in(1,\infty)$, can be approximated in the norm of $D_0^{1,q}(\Omega)\cap L^r(\Omega)$ with functions from $C_0^\infty(\Omega)$, provided $\Omega$ is bounded and locally lipschitzian, *cf.* Lemma 2.5. Such a result can be deduced from Nečas (1967, Chapitre 2). For completeness, we shall give here an independent proof.

The following result ensures the existence of a suitable "cut-off" function.

**Lemma 2.3.** *Let $\Omega$ be an arbitrary domain and put*

$$\delta(x)=\text{dist}(x,\partial\Omega). \tag{2.17}$$

*For any $\varepsilon>0$, set*

$$\gamma(\varepsilon)=\exp(-1/\varepsilon).$$

*Then, there exists a function* $\psi_\varepsilon \in C^\infty(\bar\Omega)$ *such that*

$$(i)\ |\psi_\varepsilon(x)| \le 1, \text{ for all } x\in\Omega;$$

$$(ii)\ \psi_\varepsilon(x) = 1, \text{ if } \delta(x)<\gamma^2/2\kappa_1;$$

$$(iii)\ \psi_\varepsilon(x) = 0, \text{ if } \delta(x)\ge 2\gamma(\varepsilon);$$

$$(iv)\ |D\psi_\varepsilon(x)| \le \kappa_2\varepsilon/\delta(x), \text{ for all } x\in\Omega,$$

*where* $\kappa_1$ *and* $\kappa_2$ *are positive constant independent of* x.

*Proof*. Consider the following function of $\mathbb{R}$ into itself:

$$\varphi_\varepsilon(t) = \begin{cases} 1 & \text{if } t<\gamma^2(\varepsilon) \\ \varepsilon \ln(\gamma(\varepsilon)/t) & \text{if } \gamma^2(\varepsilon)<t<\gamma(\varepsilon) \\ 0 & \text{if } t>\gamma(\varepsilon). \end{cases}$$

Clearly, choosing $\eta=\gamma^2/2$, the mollifier $\Phi_\varepsilon \equiv (\varphi_\varepsilon)_\eta$ of $\varphi_\varepsilon$ verifies $\Phi_\varepsilon(t)=1$ for $t<\gamma^2/2$, $\Phi_\varepsilon(t)=0$ for $t>2\gamma$ and

$$|\Phi'_\varepsilon(t)| \le \varepsilon/t, \quad \text{for all } t\in\mathbb{R}. \tag{2.18}$$

In addition, $|\Phi_\varepsilon(t)|\le 1$. Let

$$\psi_\varepsilon(x) \equiv \Phi_\varepsilon(\rho(x)),$$

where $\rho$ is the regularized distance of Stein (1970, Chapter VI, Theorem 2), namely, $\rho\in C^\infty(\Omega)$ and, for all $x\in\Omega$,

$$(i)\ \delta(x) \le \rho(x) \le \kappa_1\delta(x),$$

$$(ii)\ |\nabla\rho(x)| \le \kappa_2\ ,$$

with $\kappa_1$, $\kappa_2>0$. We then find

$$\psi_\varepsilon(x) = 1 \quad \text{if } \delta(x)<\gamma^2/2\kappa_1$$

$$\psi_\varepsilon(x) = 0 \quad \text{if } \delta(x)>2\gamma.$$

Moreover, from (2.18) and from properties (*i*), (*ii*) of $\rho$, it follows

$$|\nabla\psi_\varepsilon(x)| \le \kappa_2\varepsilon/\rho(x) \le \kappa_2\varepsilon/\delta(x), \text{ for all } x\in\Omega.$$

The result is therefore completely proved.

The result stated in the following lemma is well known. For a proof, we refer the reader to Nečas (1967, Théorème 2.3 at p.286)

**Lemma 2.4.** *Let* $\Omega$ *be a bounded, locally lipschitzian domain. Then there exists positive* $c=c(\Omega)$ *such that for all* $u\in W_0^{1,q}(\Omega)$, $1<q<\infty$,

$$\|u\delta^{-1}\|_q \le c|u|_{1,q}$$

*where* $\delta$ *is given in* (2.17).

With the aid of Lemmas 2.3 and 2.4 we can prove the following

**Lemma 2.5.** *Let* $\Omega$ *be as in* Lemma 2.4. *Suppose*

$$u\in D_0^{1,q}(\Omega)\cap[\bigcap_{i=1}^{k} L^{r_i}(\Omega)],$$

*with* $q,r_i\in(1,\infty)$, $i=1,\dots,k$. *Then, for any* $\eta>0$ *there is* $u_\eta\in C_0^\infty(\Omega)$ *such that*

$$|u-u_\eta|_{1,q}+\sum_{i=1}^{k}\|u-u_\eta\|_{r_i}<\eta.$$

*Proof.* For simplicity, we show the result for $k=1$. Given $\varepsilon>0$, we set

$$\theta_\varepsilon(x) = 1-\psi_\varepsilon(x),$$

where $\psi_\varepsilon(x)$ has been introduced in Lemma 2.3. We then have that $|\theta_\varepsilon(x)|\le 1$, $\theta_\varepsilon(x)$ vanishes in a neighbourhood $N_\varepsilon$ of $\partial\Omega$, $\theta_\varepsilon(x)=1$ for $\delta(x)\ge 2exp(-1/\varepsilon)$ and

$$|\nabla\theta_\varepsilon(x)|\le\kappa_2\varepsilon/\delta(x),\ \text{for all}\ x\in\Omega. \tag{2.19}$$

Putting

$$u_\varepsilon(x)=\theta_\varepsilon(x)u(x),$$

we at once recognize that $u_\varepsilon(x)$ is of compact support in $\Omega$ and that

$$\lim_{\varepsilon\to 0}\|u_\varepsilon-u\|_r = 0, \tag{2.20}$$

Furthermore, from (2.19) it is

$$|u_\varepsilon-u|_{1,q}\le\|(1-\theta_\varepsilon)\nabla u\|_q+\kappa_2\varepsilon\|u/\delta\|_q$$

and so, by Lemma 2.4, we obtain

$$|u_\varepsilon-u|_{1,q}\le\|(1-\theta_\varepsilon)\nabla u\|_q+c\varepsilon|u|_{1,q},$$

with c independent of u and $\varepsilon$. Thus

$$\lim_{\varepsilon\to 0}|u_\varepsilon-u|_{1,q} = 0. \tag{2.21}$$

Now, $u_\varepsilon$ can be approximated by its regularizer $(u_\varepsilon)_\rho$ in both spaces $W^{1,q}(\Omega)$ and $L^r(\Omega)$ and since -for $\rho$ small enough- $(u_\varepsilon)_\rho\in C_0^\infty(\Omega)$, the lemma follows from this and from (2.20), (2.21).

We are now in a position to prove the main result of this section.

**Theorem 2.2.** *Let* $\Omega$ *be a locally lipschitzian domain with a bounded boundary. Assume*

$$v\in D_0^{1,q}(\Omega)\cap[\bigcap_{i=1}^{k} L^{r_i}(\Omega)],$$

*for some* $q, r_i \in (1,\infty)$, $i=1,\dots,k$. *Then, there exists a sequence* $\{v_m\} \subset \mathcal{D}(\Omega)$ *such that*

$$\lim_{m\to\infty} |v_m - v|_{1,q} = \lim_{m\to\infty} \sum_{i=1}^{k} \|v_m - v\|_{r_i} = 0.$$

*Proof.* Again, for simplicity, we shall treat the case $k=1$. Consider first $\Omega$ bounded. By Lemma 2.5 there exists a sequence $\{\varphi_m\} \subset C_0^\infty(\Omega)$ satisfying

$$\lim_{m\to\infty} |\varphi_m - v|_{1,q} = \lim_{m\to\infty} \|\varphi_m - v\|_r = 0. \tag{2.22}$$

Since

$$v,\ \nabla\cdot v \in L^r(\Omega),$$

and $v$ has zero trace at the boundary, we have

$$\varphi_m - v \in H_{0,r}(\Omega).$$

In addition, it is

$$\nabla\cdot(\varphi_m - v) = \nabla\cdot\varphi_m \in C_0^\infty(\Omega),$$

and so, by Theorem 2.1 there exists $w_m \in C_0^\infty(\Omega)$ such that, for all $m\in\mathbb{N}$,

$$\nabla\cdot w_m = -\nabla\cdot\varphi_m$$

$$\|w_m\|_{1,q} \le c\|\nabla\cdot\varphi_m\|_q$$

$$\|w_m\|_r \le c\|\varphi_m - v\|_r$$

where $c$ is independent of $w_m$, $\varphi_m$ and $v$. Setting

$$v_m = \varphi_m + w_m,$$

there follows

$$|v - v_m|_{1,q} \le |\varphi_m - v|_{1,q} + \|w_m\|_{1,q} \le |\varphi_m - v|_{1,q} + c\|\nabla\cdot\varphi_m\|_q$$

$$\|v - v_m\|_r \le \|\varphi_m - v\|_r + \|w_m\|_r \le (1+c)\|\varphi_m - v\|_r.$$

Taking into account that $\lim_{m\to\infty} \|\nabla\cdot\varphi_m\|_q = 0$ (since $\nabla\cdot v = 0$), from these latter inequalities and from (2.22) we complete the proof of the theorem in the case $\Omega$ bounded. Assume now $\Omega$ an exterior domain and denote by $\zeta_R \in C_0^\infty(\mathbb{R}^n)$ a "cut-off" function which equals one in $\Omega_R$ and is zero in $\Omega^{2R}$ with

$$|\nabla\zeta_R| \le c_1/R, \tag{2.23}$$

with $c_1$ independent of $R$. Being

$$\int_{\Omega_{R,2R}} \nabla\cdot(\zeta_R v) = 0, \quad \|\nabla\cdot(\zeta_R v)\|_q + \|\nabla\cdot(\zeta_R v)\|_r < \infty,$$

the problem

$$\begin{aligned} &\nabla\cdot w_R = -\nabla\cdot(\zeta_R v)\\ &w_R \in W_0^{1,q}(\Omega_{R,2R})\cap W_0^{1,r}(\Omega_{R,2R}) \qquad\qquad (2.24)\\ &|w_R|_{1,s,\Omega_{R,2R}} \le c_2\|\nabla\zeta_R\cdot v\|_s\ ,\ s=q,r. \end{aligned}$$

admits a solution, *cf*. Bogovskiǐ (1980, Theorem 1) and, furthermore, the constant $c_2$ entering the estimate can be taken independent of R (this latter statement being obvious). In view of (2.23) and $(2.24)_3$, we also have

$$|w_R|_{1,s,\Omega_{R,2R}} \le c_3 R^{-1}\|v\|_{s,\Omega_{R,2R}}\ ,\ s=q,r \qquad (2.25)$$

where, again, $c_3$ does not depend on R. Set

$$v_R = \zeta_R v + w_R.$$

Clearly,

$$v_R \in D_0^{1,q}(\Omega_{2R})\cap L^r(\Omega_{2R}),\quad \nabla\cdot v_R=0.$$

Since $\Omega_{2R}$ is locally lipschitzian, it follows

$$v_R \in \mathcal{D}_0^{1,q}(\Omega_{2R})\cap L^r(\Omega_{2R}).$$

By the first part of the proof we may then state that for any $\varepsilon>0$ there is $v_{\varepsilon,R}\in\mathcal{D}(\Omega)$ such that

$$|v_R-v_{\varepsilon,R}|_{1,q} + \|v_R-v_{\varepsilon,R}\|_r < \varepsilon$$

and so

$$\|v-v_{\varepsilon,R}\|_r \le \|v_R-v_{\varepsilon,R}\|_r + \|v-v_R\|_r < \varepsilon + \|(1-\zeta_R)v\|_r + \|w_R\|_{r,\Omega_{R,2R}}. \qquad (2.26)$$

Obviously, for all sufficiently large R it is

$$\|(1-\zeta_R)v\|_r < \varepsilon. \qquad (2.27)$$

Moreover, from inequalities (1.3), (1.3)′ and (2.25) it follows

$$\|w_R\|_{r,\Omega_{R,2R}} \le c_4 R|w_R|_{1,r,\Omega_{R,2R}} \le c_5\|v\|_{r,\Omega_{R,2R}}$$

with c independent of R and so, again for all sufficiently large R,

$$\|w_R\|_{s,\Omega_{R,2R}} < \varepsilon. \qquad (2.28)$$

From (2.25)-(2.28) we thus find that $v_{\varepsilon,R}\subset\mathcal{D}(\Omega)$ approaches $v$ in $L^r$. It remains to show that $v_{\varepsilon,R}$ approaches $v$ also in $D^{1,q}$. Reasoning as before, we prove

$$|v-v_{\varepsilon,R}|_{1,q} < 2\varepsilon + |w_R|_{1,q,\Omega_{R,2R}} < 2\varepsilon + c_3 R^{-1}\|v\|_{q,\Omega_{R,2R}}, \tag{2.29}$$

and so the proof of the theorem follows from (2.29) once we establish that

$$R^{-1}\|v\|_{q,\Omega_{R,2R}} < \varepsilon \tag{2.30}$$

for all sufficiently large R. Now, if $q\in(1,3)$, by Hölder inequality it is

$$\|v\|_{q,\Omega_{R,2R}} \le c_1 R\|v\|_{3q/(3-q),\Omega^R},$$

with $c_6=c_6(q)$ and since, by Sobolev inequality,

$$\|v\|_{3q/(3-q),\Omega} < \infty,$$

we conclude (2.30). If $r\ge q\ge 3$ the same conclusion holds since, in such a case, by Hölder inequality,

$$R^{-1}\|v\|_{q,\Omega_{R,2R}} \le c_2 R^{n(1/q-1/r-1/3)}\|v\|_{r,\Omega_{R,2R}},$$

and (2.30) is recovered. Finally, assume $q\in[3,\infty)\cap(r,\infty)$. We recall a result due to Nirenberg (1959, Theorem at p.125), which states that, if $u\in C_0^\infty(\mathbb{R}^3)$ then the following inequality holds

$$\|u\|_q \le c|u|_{1,q}^{\lambda}\|u\|_r^{1-\lambda}, \tag{2.31}$$

with

$$\lambda = \frac{3(q-r)}{r(q-3)+3q} \qquad (<1)$$

and $c=c(q,r)$. By Theorem 1.2 and by (2.31), we thus recover that every function in $D^{1,q}(\mathbb{R}^n)\cap L^r(\mathbb{R}^n)$, $q\in[3,\infty)\cap(r,\infty)$, belongs, in fact, to $L^q(\mathbb{R}^3)$. Since $v\in D^{1,q}(\mathbb{R}^n)\cap L^r(\mathbb{R}^n)$, we then find $v\in L^q(\mathbb{R}^n)$ and (2.30) again follows. The proof is therefore completed.

## 3. Existence, Uniqueness and $L^q$-estimates for the Oseen Problem in an Exterior Domain.

In this section we recall some results concerning the solutions to the Oseen problem in a three-dimensional exterior domains. They are a special case of those recently obtained by Galdi (1992a).

Let us consider the following Oseen problem:

$$\left.\begin{aligned}\Delta u-\lambda\frac{\partial u}{\partial x} &= \nabla\pi+F\\ \nabla\cdot u&=0\end{aligned}\right\}\ \text{in } \Omega \tag{3.1}$$
$$\lim_{|x|\to\infty} u(x) = 0,$$
$$u = u_* \ \text{at } \partial\Omega$$

where $\lambda>0$ and $F$ and $u_*$ are prescribed functions in $\Omega$ and on $\partial\Omega$, respectively. The following result establishes existence of solutions to (3.1) and validity of corresponding estimates.

**Theorem 3.1.** *Let $\Omega$ be an exterior domain in $\mathbb{R}^3$ of class $C^2$. Given*

$$F\in L^q(\Omega),\ u_*\in W^{2-1/q,q}(\partial\Omega),\ 1<q<2,$$

*there exists at least one corresponding pair $u,\pi$ which solves the Oseen problem (3.1) a.e. in $\Omega$ and such that*

$$u\in L^{s_2}(\Omega)\cap D^{1,s_1}(\Omega)\cap D^{2,q}(\Omega),\ \pi\in D^{1,q}(\Omega)$$

*with $s_1=4q/(4-q)$, $s_2=2q/(2-q)$. Moreover, $u,\pi$ verify the estimate*

$$\alpha_1\|u\|_{s_2}+\lambda\left\|\frac{\partial u}{\partial x_1}\right\|_q+\alpha_2|u|_{1,s_1}+|u|_{2,q}+|\pi|_{1,q}\le c(\|F\|_q+\|u_*\|_{2-1/q,q(\partial\Omega)}) \tag{3.2}$$

*where*

$$\alpha_1 = \min\{1,\lambda^{1/2}\},\ \alpha_2 = \min\{1,\lambda^{1/4}\}.$$

*The constant c depends on $q,\Omega$ and $\lambda$. However, if $q\in(1,3/2)$ and $\lambda\in(0,B]$ for some $B>0$, c depends solely on $q,\Omega$ and $B$.*

We shall now consider the uniqueness of solutions of Theorem 3.1. This problem can be easily solved in the same class of existence, *cf.* Galdi (1992a). However, in the present context, we need to establish uniqueness of such solutions in an *a priori* larger class (as far as the behaviour at infinity is concerned). Specifically, we have

**Theorem 3.2.** *Let $\Omega$, $F$, $u_*$, $u$ and $\pi$ be as in* Theorem 3.1. *Let $u_1,\pi_1$ be another solution to* (3.1) *corresponding to $F$ and $u_*$ and such that $u_1\in W^{2,q}(\Omega_R)$, $\pi_1\in L^q(\Omega_R)$, for all $R>\delta(\Omega^c)$*[6]*. Then, if $u_1\in D^{1,2}(\Omega)$, it follows $u=u_1$ and $\pi=\pi_1$+const. a.e. in $\Omega$.*

*Proof.* Setting $w=u_2-u_1$, $\tau=\pi_2-\pi_1$, from (3.1) we deduce

---

[6] The assumptions on $u_1$ and $\pi_1$ can be fairly weakened. However, this would be unessential for further porposes.

$$\left.\begin{array}{l}\Delta \mathbf{w} - \lambda \dfrac{\partial \mathbf{w}}{\partial x_1} = \nabla \tau \\ \nabla \cdot \mathbf{w} = 0\end{array}\right\} \text{ in } \Omega$$
$$\lim_{|x|\to\infty} \mathbf{w}(x) = 0, \qquad (3.3)$$
$$\mathbf{w} = 0 \quad \text{at } \partial\Omega$$

Well-known elliptic theory ensures $\mathbf{w},\tau \in C^{\infty}(\Omega)$. Moreover, from Theorem 1.1, we have

$$u_1 \in L^{3q/(3-q)}(\Omega), \ u_2(\Omega) \in L^{6}(\Omega).$$

Thus,

$$\int_\Omega \mathbf{w}^2(x)(1+|x|)^{-5} < \infty \ ,$$

and from Theorem 4 of Chang & Finn (1961) we find

$$\mathbf{w}(x) = O(|x|^{-1}),$$

which, in turn, by Theorem 6 of Chang & Finn (1961), furnishes $\mathbf{w} \equiv 0$ in $\Omega$. The proof is thus completed.

## 4. *D*-Solutions and Related Results.

We begin to recall the definition of *D*-solution. Specifically, we have

**Definition 4.1.** A vector field $\mathbf{v}:\Omega \to \mathbb{R}^3$ is called a *D-solution to the Navier-Stokes problem* (0.1), (0.2) if and only if

*(i)* $\mathbf{v} \in D^{1,2}(\Omega)$;

*(ii)* $\nabla \cdot \mathbf{v} = 0$;

*(iii)* $\mathbf{v}$ satisfies the boundary condition $(0.2)_1$ (in the trace sense) or, if $\mathbf{v}_* \equiv 0$, then $\vartheta \mathbf{v} \in D_0^{1,2}(\Omega)$ where $\vartheta \in C_0^1(\mathbb{R}^3)$ and $\vartheta(x)=1$ if $x \in \Omega_{R/2}$ and $\vartheta(x)=0$ if $x \in \Omega^R$, $R > 2\delta(\Omega^c)$;

*(iv)* $\displaystyle \lim_{|x|\to\infty} \int_S |\mathbf{v}(x) - \mathbf{v}_\infty| = 0$;

*(v)* $\mathbf{v}$ satisfies the identity

$$(\nabla \mathbf{v}, \nabla \varphi) + (\mathbf{v} \cdot \nabla \mathbf{v}, \varphi) = -\,[\mathbf{f}, \varphi] \qquad (4.1)$$

for all $\varphi \in \mathcal{D}(\Omega)$.

**Remark 4.1.** If the body force $f$ belongs to $W^{-1,2}(\Omega')$ for every bounded domain $\Omega'$ with $\overline{\Omega}'\subset\Omega$, to any *D*-solution we can associate a pressure field $p\in L^2_{loc}(\Omega)$ by the following argument. Consider the functional

$$\mathcal{F}(\psi)=(\nabla v,\nabla\psi) + (v\cdot\nabla v,\psi) + [f,\psi],\ \psi\in D_0^{1,2}(\Omega').$$

Since $v\in W^{1,2}_{loc}(\Omega)$, *cf*. Nečas (1967, p.115, Exercice 7.1), $\mathcal{F}$ is bounded in $D_0^{1,2}(\Omega')$ and vanishes in $\mathcal{D}_0^{1,2}(\Omega')$. Then, by the arbitrarity of $\Omega'$, from a result of Solonnikov & Ščadilov (1973), it follows the existence of $p\in L^2_{loc}(\Omega)$ such that

$$(\nabla v,\nabla\psi) + (v\cdot\nabla v,\psi) + [f,\psi] = (p,\nabla\cdot\psi),\quad \text{for all } \psi\in C_0^\infty(\Omega). \tag{4.2}$$

If, in particular, $\Omega$ is locally lipschitzian, one has $p\in L^2(\Omega_R)$, for all $R>\delta(\Omega^c)$.

**Remark 4.2.** If $f\in D^{-1,2}(\Omega)\cap D^{-1,3}(\Omega)$, then the pressure field of a *D*-solution corresponding to $f$ belongs to $L^3(\Omega)$, *cf*. Galdi (forthcoming).

We shall now recall a classical result on the existence of *D*-solutions. To this end, for given $v_*\in W^{1/2,2}(\partial\Omega)$ and $v_\infty\in\mathbb{R}^3$, we shall say that $V$ is *an extension of* $v_*$ *and* $v_\infty$ if and only if for some $\rho>\delta(\Omega^c)$

(*i*) $|V(x)-v_\infty|\le c|x|^{-n+1}$, for all $x\in\Omega^\rho$;

(*ii*) $V\in D^{1,s}(\Omega^\rho)\cap W^{1,2}(\Omega_\rho)$, for all $s>1$;

(*iii*) $V=v_*$ at $\partial\Omega$;

(*iv*) $\nabla\cdot V=0$ in $\Omega$

The following result can be deduced from the work of Fujita (1961) and Finn (1965), *cf*. Galdi (forthcoming).

**Theorem 4.1.** *Given*

$$f\in D^{-1,2}(\Omega),\ v_*\in W^{1/2,2}(\partial\Omega),\ v_\infty\in\mathbb{R}^3,$$

*then, there is $M>0$ such that if*

$$\left|\int_{\partial\Omega} v_*\cdot n\right| < M,$$

*the Navier-Stokes problem* (0.1), (0.2) *admits at least one D-solution. This solution satisfies the inequality*

$$|v|_{1,2}+[f,v-V] \le (\nabla v,\nabla V)-(v\cdot\nabla V,v-V), \tag{4.3}$$

*where* $V$ *is an extension of* $v_*$ *and* $v_\infty$. *In addition, if* $f$ *is such that*

$$f\in L^q(\Omega^R),\ \textit{for some}\ R>\delta(\Omega^c)\ \textit{and}\ q\in(3/2,\infty),$$

*then*

$$\lim_{|x|\to\infty} v(x)=v_\infty,$$

*uniformly.*

**Remark 4.2.** Relation (4.3) is the so called *energy inequality*. This because, if $v$ is smooth in $\Omega_R$ for some $R>\delta(\Omega^c)$, (4.3) reduces to (0.3) with the equality sign replaced by the inequality sign, *cf.* also Section 5. On the other hand, the inequality sign is not due to a lack of differentiability of the solution but, rather, to a lack of information regarding *its behaviour at large distances*. Actually, a *D*-solution in a *bounded* domain[7] always verifies an energy *equality*. It is interesting to compare this situation with that concerning weak (Hopf) solutions to the *time dependent* Navier-Stokes equations in a *bounded* domain where only an energy inequality is known to hold, *cf.* Hopf (1951).

The differentiability of *D*-solutions is established in the next theorem, *cf.* Temam (1977).

**Theorem 4.2.** *Let* $v$ *be a D-solution to* (0.1), (0.2) *in an exterior domain* $\Omega\subseteq\mathbb{R}^3$, *of class* $C^{m+2}$. *Then, if for some* $q>1$,

$$f\in W^{m,q}(\Omega_R),\ v_*\in W^{m+2-1/q,q}(\partial\Omega),$$

*it follows*

$$v\in W^{m+2,q}(\Omega_R),\quad p\in W^{m+1,q}(\Omega_R).$$

*Further, if* $\Omega$ *is of class* $C^\infty$ *and*

$$f\in C^\infty(\bar{\Omega}_R),\ v_*\in C^\infty(\partial\Omega)$$

*then*

$$v,p\in C^\infty(\bar{\Omega}_R).$$

---

[7] Namely, a vector field $V$ satisfying (i), (ii) and (iv) of Definition 1.1.

## 5. On the Validity of the Energy Equation for *D*-solutions.

The objective of the present section is to investigate under which conditions on the data a corresponding *D*-solution satisfies the energy equation. As already observed in the introduction and in Remark 4.2, this problem arises only for flows occurring in an unbounded region $\Omega$, since, for $\Omega$ bounded, every such a solution obeys the above equation. Actually, *for* $\Omega$ bounded, the *proof of* the energy equation (*i.e.* the relation formally obtained from (0.3) with $v_\infty=0$) depends *solely* on the *regularity in* $\bar{\Omega}$ of the solution and one readily shows that, if $f$ has a mild degree of smoothness, every corresponding *D*-solution satisfies the energy equation. However, if $\Omega$ *is* an exterior domain, the proof of (0.1) demands not only certain regularity in $\bar{\Omega}_R$, for all $R>\delta(\Omega^c)$, but it depends also in an essential way on the *asymptotic properties* of the solution. The fact that not every solution *merely* having finite Dirichlet integral is expected to satisfy the energy equation in an exterior domain can be easily guessed via the following arguments. Assume $v,p$ is a solution to (0.1), (0.2). For simplicity, we shall assume that $v,p$ is classical and that $v_*\equiv 0$. Multiplying $(0.1)_1$ by $u\equiv v-v_\infty$, integrating by parts over $\Omega_R$ and taking into account $(0.1)_2$, we deduce

$$\int_{\Omega_R} \nabla v:\nabla v + v_\infty\cdot\int_{\partial\Omega} T(v,p)\cdot n + \int_{\Omega_R} f\cdot(v-v_\infty)$$
$$= \int_{\partial B_R} u\cdot T(v,p)\cdot n - \frac{1}{2}\int_{\partial B_R} u^2 v\cdot n \,. \tag{5.1}$$

If $v$ is a *D*-solution, it is

$$v\in D^{1,2}(\Omega) \tag{5.2}$$

and so if $f$ is "well-behaved" at infinity, letting $R\to\infty$ into (5.1) we have that the left hand side of (5.1) tends to the left hand side of (0.3) (with $v_*=0$); nevertheless the *sole* condition (5.2) together with the condition $p\in L^3(\Omega)$, *cf*. Remark 4.2, is *not* enough to ensure that the surface integral at the right hand side of (5.1) tends to zero as $R\to\infty$, even along a sequence. Thus, we are not able to conclude the validity of (0.3), unless we have *further* information on the behaviour of $v$ at large distances.

The reasonings just developed well explain why, in the case of an exterior domain $\Omega$, a *D*-solution should be still considered a "generalized" solution even though it is smooth in $\bar{\Omega}_R$, for any $R>\delta(\Omega^c)$. Actually, unlike the case $\Omega$ bounded where the word "generalized" expresses only a possible lack of differentiability, for $\Omega$ exterior it is also to mean the possibility that the solution need not be "regular" in the neighbourhood of infinity as expected from the physical (and intuitive) point of view.

We shall presently investigate what regularity must be imposed on a *D*-solution at large distances in order that it satisfies the equation (0.3). We shall first consider the case $v_*\equiv 0$ and denote the corresponding energy equation (0.3) by $(0.3)_0$.

In order to perform this investigation, however, there is, preliminarily, a formal aspect which has to be fixed up. Actually, if $v$ merely belongs to $W^{1,2}_{loc}(\Omega_R)$ for all $R>\delta(\Omega^c)$ –as prescribed by Definition 4.1– the second integral in $(0.3)_0$ need not be meaningful. Thus, in such a case, we have to introduce a suitable generalization of $(0.3)_0$. Let us denote by $a$ any vector field in $\Omega$ verifying

$$
\begin{aligned}
&(i)\ a\in C^\infty(\Omega);\\
&(ii)\ \nabla\cdot a = 0 \text{ in } \Omega; \qquad\qquad (5.3)\\
&(iii)\ a=0 \text{ in } \Omega_d \text{ and } a=v_\infty \text{ in } \Omega^{2d} \text{ for some } d>\delta(\Omega^c).
\end{aligned}
$$

The field $a$ will be called *an extension of* $v_\infty$. For instance, we may take

$$a = v_\infty-\Delta(\zeta b) +\nabla[\nabla\cdot(\zeta b)]$$

where $\zeta$ is an arbitrary $C^\infty$ function in $\bar{\Omega}$ which is one near $\partial\Omega$ and is zero far from $\partial\Omega$, while

$$b = \frac{1}{2}\, v_\infty x_2^2.$$

We then give the following

**Definition 5.1.** Let $v$ be a *D*-solution to (0.1), (0.2) corresponding to $v_*\equiv 0$ and let $a$ be an extension of $v_\infty$. The relation

$$|v|_{1,2} +[f,v-a] = (\nabla v,\nabla a) -(v\cdot\nabla a,v-a) \qquad (5.4)$$

is called *generalized energy equation.*

To justify the above definition, we notice that (5.4) reduces to $(0.3)_0$ provided $\Omega$, $f$, $v$ and $p$ are sufficiently smooth. For example, if $\Omega$ is of class $C^1$ and

$$v\in C^1(\bar{\Omega}_R)\cap C^2(\Omega_R),\ p\in C(\bar{\Omega}_R)\cap C^1(\Omega_R),\ f\in C(\Omega),$$

multiplying (0.1) by $a-v_\infty$, integrating by parts over $\Omega_R$ and observing that *supp*$(\nabla a)\subset\bar{\Omega}_{d,2d}$ there follows

$$-\int_{\partial\Omega} v_\infty\cdot T(v,p)\cdot n = [f,a-v_\infty] +(\nabla v,\nabla a) -(v\cdot\nabla a,v-a). \qquad (5.4)'$$

This identity along with (5.4) implies $(0.3)_0$.

**Remark 5.1** The validity of (5.4)′ can be shown under very mild regularity assumptions on f. For example, suppose that, for some $R>\delta(\Omega^c)$, $f\in L^r(\Omega_R)$ with $r\in[3/2,\infty)$. Assume, further, $v_*\equiv 0$ and $\Omega$ of class $C^2$. Then, from Theorem 4.2 it easily follows that every *D*-solution v corresponding to these data satisfies (5.4)′ with p pressure field associated to v (*cf.* Remark 4.1).

**Remark 5.2.** As we know from Theorem 4.1, for any $f\in D_0^{-1,2}(\Omega)$ there exists a corresponding generalized solution which (for $v_*\equiv 0$) satisfies the *generalized energy inequality*, that is, relation (5.4) with "=" replaced by "≤".

Our next objective is to prove a quite general result which furnishes sufficient conditions under which a *D*-solution obeys the generalized energy equation, *cf.* Lemma 5.2. Successively, we shall prove that, if the data f and $v_*$ satisfy certain requirements of summability, then every corresponding *D*-solution meets the above said conditions and, therefore, it satisfies the generalized energy equation. Actually, the assumptions on the data are such that the equation is satisfied in its classical form (0.3), *cf.* Theorem 5.1. As a corollary to this result, we deduce a theorem of Liouville type, *cf.* Theorem 5.2.

To show all this, we first need a rather simple but useful result which we state in the form of a lemma.

**Lemma 5.1.** *Let $\Omega$ be an arbitrary domain in $R^3$. Then the trilinear form*

$$a(u,v,w) \equiv (u\cdot\nabla v,w) \qquad (5.5)$$

*is continuous in the space*

$$L^q(\Omega)\times D^{1,r}(\Omega)\times L^s(\Omega), \quad q^{-1}+r^{-1}+s^{-1}=1.$$

*and in the space*

$$D_0^{1,2}(\Omega)\times D^{1,2}(\Omega)\times L^3(\Omega).$$

*Proof.* The first assertion is a trivial consequence of the Hölder inequality, while the second one is proved by choosing q=6, r=3 and using the Sobolev inequality (1.4)

$$\|u\|_6 \le (2/\sqrt{3})\,|u|_{1,2}.$$

We are now in a position to prove the following

**Lemma 5.2.** *Let* $v$ *be a D-solution to the Navier-Stokes problem* (0.1), (0.2) *in an exterior, locally lipschitzian domain, corresponding to* $f\in D_0^{-1,2}(\Omega)$, $v_*=0$ *and to some* $v_\infty\in\mathbb{R}^3-\{0\}$. *Then, a sufficient condition in order that* $v$ *satisfies the generalized energy equation* (5.4) *for any extension* a *of* $v_\infty$ *is*

$$\begin{aligned} &v-v_\infty\in L^4(\Omega)\cap L^q(\Omega)\\ &v_\infty\cdot\nabla v\in L^{q'}(\Omega)\end{aligned} \qquad (5.6)$$

*for some* $q\in(1,\infty)$, $q'=q/(q-1)$.

*Proof.* Let a be any extension of $v_\infty$ and set

$$u=v-a.$$

Since $\Omega$ is locally lipschitzian, by *(i)-(iv)* of Definition 4.1 and in view of the characterization (1.1), we obtain

$$u\in\mathcal{D}_0^{1,2}(\Omega). \qquad (5.7)$$

Furthermore, from $(5.6)_1$ we also have

$$u\in L^4(\Omega)\cap L^q(\Omega). \qquad (5.8)$$

By (5.7) and (5.8), we may employ Theorem 2.2 to deduce the existence of a sequence $\{u_k\}\subset\mathcal{D}(\Omega)$ converging to u in the space

$$\mathcal{D}_0^{1,2}(\Omega)\cap L^4(\Omega)\cap L^q(\Omega).$$

Setting $\varphi=u_k$ into (4.1) we thus find

$$(\nabla v,\nabla u_k)+((v-v_\infty)\cdot\nabla v,u_k)+(v_\infty\cdot\nabla v,u_k)=-\mathcal{R}e[f,u_k]. \qquad (5.9)$$

Taking into account Lemma 5.1, (5.6)-(5.8), we recover, in the limit $k\to\infty$

$$\begin{aligned} &(\nabla v,\nabla u_k)\longrightarrow(\nabla v,\nabla u)\\ &((v-v_\infty)\cdot\nabla v,u_k)\longrightarrow((v-v_\infty)\cdot\nabla v,u)\\ &(v_\infty\cdot\nabla v,u_k)\longrightarrow(v_\infty\cdot\nabla v,u).\end{aligned} \qquad (5.10)$$

Moreover, since for all $k\in\mathbb{N}$ it is

$$((v-v_\infty)\cdot\nabla u_k,u_k) = 0,$$

we have, again by Lemma 5.1,

$$((v-v_\infty)\cdot\nabla u,u) = 0$$

and (5.9), (5.10) furnish

$$|v|_{1,2}+[f,v-a] = (\nabla v,\nabla a)-(v\cdot\nabla a,v-a)+(v_\infty\cdot\nabla u,u). \tag{5.11}$$

We have now to show that the last term at the right hand side of (5.11) vanishes identically. First of all, we observe that it is clearly well-defined since, by $(5.6)_2$ and (5.8) it follows

$$v_\infty\cdot\nabla u\cdot u \in L^1(\Omega). \tag{5.12}$$

Next, taking $v_\infty=\lambda e_1$, for $\rho>\delta(\Omega^c)$ we let $\{\mathscr{C}_{k,\rho}\}$, $k\in\mathbb{N}$, denote a family of cylinder-like domains of the form

$$\mathscr{C}_{k,\rho} = \left\{x\in\Omega:\ |x'|<k,\ |x_1|<\rho\right\},$$

where $x'=(x_2,x_3)$. For sufficiently large k and $\rho$ it is $\mathscr{C}_{k,\rho}\supset\Omega^c$ and, correspondingly, we find

$$\int_{\mathscr{C}_{k,\rho}} v_\infty\cdot\nabla u\cdot u = \lambda\int_{\mathscr{C}_{k,\rho}}\frac{\partial u}{\partial x_1}\cdot u = \lambda\int_{\Sigma_1(k,\rho)} u^2 - \lambda\int_{\Sigma_2(k,\rho)} u^2, \tag{5.13}$$

where

$$\Sigma_1(k,\rho) = \left\{x\in\mathbb{R}:\ x_1=\rho,\ |x'|\le k\right\},$$

$$\Sigma_2(k,\rho) = \left\{x\in\mathbb{R}:\ x_1=-\rho,\ |x'|\le k\right\}.$$

By $(5.6)_1$ and (5.7), it is easy to show that, for each fixed k,

$$\lim_{\rho\to\infty}\int_{\Sigma_i(k,\rho)} u^2 = 0, \quad i=1,2. \tag{5.14}$$

In fact, setting $x_\rho=(\rho,0,0)$, by a well known boundary inequality it follows

$$\left(\int_{\Sigma_1(k,\rho)} u^2\right)^{\frac{1}{2}} \le c_1\left(\|u\|_{2,B_k(x_\rho)}+\|\nabla u\|_{2,B_k(x_\rho)}\right)$$

$$\le c_2\left(\|u\|_{4,B_k(x_\rho)}+\|\nabla u\|_{2,B_k(x_\rho)}\right),$$

with $c_1=c_1(k)$, and so (5.14) with i=1 is recovered. By an analogous argument one shows (5.14) with i=2. Putting

$$\mathscr{C}_k = \left\{x\in\Omega:\ |x'|<k,\ x_1\in\mathbb{R}\right\},$$

from (5.12), (5.13) and (5.14) we find

$$\int_{\mathscr{C}_k} v_\infty\cdot\nabla u\cdot u = 0$$

for all sufficiently large k. On the other hand, again by (5.14) it is

$$\lim_{k\to\infty}\int_{\mathscr{C}_k} v_\infty\cdot\nabla u\cdot u = \int_\Omega v_\infty\cdot\nabla u\cdot u$$

and so

$$(v_\infty\cdot\nabla u,u) = 0,$$

which completes the proof of the theorem.

**Remark 5.3.** The theorem continues to hold under different assumptions on $f$. In fact, $f$ is required to satisfy

$$[f,v_k-a] \longrightarrow [f,v-a].$$

Therefore, one can take

$$f\in L^{4/3}(\Omega),$$

or else,

$$f\in L^{q'}(\Omega).$$

Lemma 5.2 can be extended, with no essential technical changes, to the case when the velocity $v_*$ at the boundary is not identically zero. To this end, we need a suitable extension of *both* $v_*$ and $v_\infty$. For $\Omega$ locally lipschitzian and

$$v_*\in W^{1/2,2}(\partial\Omega),$$

we may take

$$A = V + m\nabla(1/4\pi|x|) + a. \tag{5.15}$$

Here $V\in W^{1,2}(\Omega)$ is an extension of $v_*-m\nabla(1/4\pi|x|)$ vanishing in $\Omega^\rho$ for some $\rho>\delta(\Omega^c)$, $m$ is the flux of $v_*$ through $\partial\Omega$ and a is an extension of $v_\infty$ in the sense of (5.3). From (5.15) it is apparent that

$$\begin{aligned}&(i)\ |A(x)-v_\infty|\le cm|x|^{-2} \quad \text{in } \Omega^\rho;\\ &(ii)\ A\in D^{1,s}(\Omega^\rho)\cap W^{1,2}(\Omega_\rho)\cap C^\infty(\Omega^\rho), \quad \text{for all } s>1;\\ &(iii)\ A=v_* \quad \text{at } \partial\Omega;\\ &(iv)\ \nabla\cdot A=0 \quad \text{in } \Omega.\end{aligned} \tag{5.16}$$

In particular, if $m=0$, $A-v_\infty$ is of bounded support in $\Omega$. Any field $A$ satisfying conditions *(i)–(iv)* in (5.16) will be called *an extension of* $v_*$ *and* $v_\infty$. Thus, generalizing Definition 5.1, we give

**Definition 5.2.** Let $v$ be a *D*-solution to (0.1), (0.2) and let $A$ be an extension of $v_*$ and $v_\infty$. The relation

$$|v|_{1,2}+[f,v-A] = (\nabla v,\nabla A) -(v\cdot\nabla A,v-A) \tag{5.17}$$

is called *generalized energy equation.*

**Remark 5.4.** As in the case $v_*=0$, under suitable regularity assumptions on the data and on $v$, one easily shows that the identity (5.17) reduces to the energy equation in its classical formulation (0.3). Specifically, assuming $\Omega$ of class $C^2$ and

$$f\in L^r(\Omega_R)\cap D_0^{-1,2}(\Omega),\ v_*\in W^{2-1/r,r}(\partial\Omega),\ \text{some } R>\delta(\Omega^c) \text{ and } r\in[3/2,\infty),$$

by an easily justified integration by parts which rests on Theorem 4.2 we find

$$\int_{\partial\Omega}(v-v_\infty)\cdot T(v,p)\cdot n -\frac{1}{2}\int_{\partial\Omega}(v_*-v_\infty)^2 v_*\cdot n = [f,A-v_\infty]+(\nabla v,\nabla A)-(v\cdot\nabla A,v-A),$$

which, in view of (5.17), implies (0.3).

By a procedure completely analogous to that used in the proof of Lemma 5.2 one can show the following result.

**Lemma 5.3.** *Let* $v$ *be a D-solution to the Navier-Stokes problem* (0.1), (0.2) *in an exterior, locally lipschitzian domain corresponding to*

$$f\in D_0^{-1,2}(\Omega),\ v_*\in W^{1/2,2}(\partial\Omega),\ v_\infty\in R^3-\{0\}.$$

*Then, a sufficient condition in order that* $v$ *satisfies* (5.17) *for any extension* $A$ *of* $v_*$ *and* $v_\infty$ *is that* (5.6) *holds for some* $q>3/2$.

**Remark 5.5.** The restriction on q (not needed in Lemma 5.2) is due to the fact that, if $m\neq 0$, the field **A** belongs to $L^q(\Omega^\rho)$ for $q>3/2$ only. However, if $m=0$, **A** can be taken of bounded support and we may assume $q\in(1,\infty)$.

**Remark 5.6.** Remark 5.3 equally applies to Lemma 5.3.

The next theorem is the main result of this section and it establishes that, under very general conditions on the data, every *D*-solution satisfies the energy equation (0.3).

**Theorem 5.1.** *Let* $\Omega$ *be a* $C^2$*-smooth, exterior domain and let* **v** *be a D-solution to the Navier Stokes problem* (0.1), (0.2) *with*

$$f\in L^{4/3}(\Omega)\cap L^{3/2}(\Omega),\ \mathbf{v}_*\in W^{4/3,3/2}(\partial\Omega). \tag{5.18}$$

*Then* **v**,p *satisfy the energy equation* (0.3), *where* p *is the pressure field associated to* **v** *by* Remark 4.1.

*Proof.* In view of Lemmas 5.2, 5.3 and of Remarks 5.5 and 5.6, it suffices to show that

$$\mathbf{v}-\mathbf{v}_\infty\in L^4(\Omega),\qquad \mathbf{v}_\infty\cdot\nabla\mathbf{v}\in L^{4/3}(\Omega). \tag{5.19}$$

Set $\mathbf{v}-\mathbf{v}_\infty=\mathbf{u}$, $\mathbf{v}_\infty=\lambda\mathbf{e}_1$, $\lambda>0$. From (5.18) and from Theorem 4.2 we then have

$$\mathbf{u}\in W^{2,3/2}(\Omega_R),p\in W^{1,3/2}(\Omega_R),\quad \text{for all } R>\delta(\Omega^c). \tag{5.20}$$

In addition, **u**,p satisfy a.e. the following Oseen problem

$$\left.\begin{array}{l}\Delta\mathbf{u}-\lambda\dfrac{\partial\mathbf{u}}{\partial x}=\nabla\pi+\mathbf{F}\\ \nabla.\mathbf{u}=0\end{array}\right\}\text{ in }\Omega$$
$$\lim_{|x|\to\infty}\mathbf{u}(x)=0,$$
$$\mathbf{u}=\mathbf{u}_*\ \text{ at } \partial\Omega$$

with $\mathbf{u}_*=\mathbf{v}_*-\lambda\mathbf{e}_1$ and $\mathbf{F}=\mathbf{f}-\mathbf{u}\cdot\nabla\mathbf{u}$. Since

$$\mathbf{u}\in D^{1,2}(\Omega), \tag{5.21}$$

from Theorem 1.1 it is

$$\mathbf{u}\in L^6(\Omega) \tag{5.22}$$

and so, from Hölder inequality and from inequality (1.5) (with q=2) there follows

$$\|u.\nabla u\|_{3/2}\leq \|u\|_6|u|_{1,2}\leq\gamma_1|u|^2_{1,2}$$

yielding

$$u\cdot\nabla u\in L^{3/2}(\Omega).$$

From this and from the assumption we then have

$$F\in L^{3/2}(\Omega).$$

Being, obviously,

$$u_*\in W^{4/3,3/2}(\partial\Omega),$$

(5.20), (5.21) and Theorems 3.1, 3.2 ensure, in particular,

$$\frac{\partial u}{\partial x_1}\in L^{3/2}(\Omega)$$
$$\left\|\frac{\partial u}{\partial x_1}\right\|_{3/2}\leq c(\|F\|_{3/2}+\|u_*\|_{4/3,3/2(\partial\Omega)}) \tag{5.23}$$

In view of (5.21) and (5.23) we may apply Theorem 1.2 with q=3/2 to deduce $u\in L^{9/2}(\Omega)$, along with the estimate

$$\|u\|_{9/2}\leq c_2(\left\|\frac{\partial u}{\partial x}\right\|_{3/2}+|u|_{1,2}) \tag{5.24}$$

As a consequence, from (5.23) we infer

$$\|u\|_{9/2}\leq c_3(\|f\|_{3/2}+1) \tag{5.25}$$

with $c_3=c_3(\Omega,v_*,\lambda,|u|_{1,2})$. A comparison between (5.22) and (5.25) shows that the exponent of summability has decreased from 6 to 9/2, implying a "better" behavior of u at large distances. After having established (5.23) and (5.24) we shall use an iterative argument which eventually leads to (5.19). Actually, for $1<q\leq3/2$, we have $u_*\in W^{2-1/q,q}(\partial\Omega)$ and, moreover,

$$\|F\|_q\leq \|f\|_q+\|u.\nabla u\|_q\leq \|u\|_{2q/(2-q)}|u|_{1,2}+\|f\|_q. \tag{5.26}$$

Therefore, repeating the reasoning previously employed, we deduce

$$\frac{\partial u}{\partial x_1}\in L^q(\Omega),$$
$$\left\|\frac{\partial u}{\partial x_1}\right\|_q\leq c_4(\|F\|_q+\|u_*\|_{4/3,3/2(\partial\Omega)}) \tag{5.27}$$

which, by (2.8) and Theorem 1.2 implies

$$u\in L^{3q}(\Omega)$$

along with the estimate

$$\|u\|_{3q}\leq c_5(\left\|\frac{\partial u}{\partial x_1}\right\|_q+|u|_{1,2}).$$

So, from (5.26) and $(5.27)_2$, we obtain

$$\|u\|_{3q}+\left\|\frac{\partial u}{\partial x_1}\right\|_q \le c_6(\|f\|_q+\|u\|_{2q/(2-q)}|u|_{1,2}+1) \tag{5.28}$$

with $c_6=c_6(\Omega,q,v_*,\lambda,|u|_{1,2})$. Let us calculate (5.28) along the sequence $\{q_k\}$ defined by the recursive formula

$$\frac{2q_{k+1}}{2-q_{k+1}}=3q_k, \qquad q_1=2. \tag{5.29}$$

We thus find, for any $k\in\mathbb{N}$,

$$\|u\|_{3q_k}+\left\|\frac{\partial u}{\partial x_1}\right\|_{q_k} \le M, \tag{5.30}$$

for a constant $M$ independent of $k$. Clearly, the sequence $\{q_k\}$ is strictly decreasing and is bounded from below by 4/3. Therefore, there exists a number $q_0\ge 4/3$ such that

$$\lim_{k\to\infty} q_k = q_0. \tag{5.31}$$

Now, passing to the limit $k\to\infty$ into (5.28) we recover

$$3q_0 = \frac{2q_0}{2-q_0}$$

namely,

$$q_0 = 4/3. \tag{5.32}$$

Property (5.19) becomes then a consequence of (5.30), (5.31) and (5.32) and the theorem is completely proved.

From Theorem 5.3 we deduce at once the following result of the Liouville type.

**Theorem 5.2.** *Let* v *be D-solution to the Navier-Stokes problem* (0.1), (0.2) *with* $\Omega=\mathbb{R}^3$ *and corresponding to* $f\equiv 0$. *Then* $v\equiv v_\infty$ *and* p≡const., *where* p *is the pressure field associated to* v *in* Remark 4.1.

**Remark 5.7.** From the proof of Theorem 5.3 it follows, in particular, that every *D*-solution v corresponding to f and $v_*$ satisfying (5.18) necessarily obeys (5.19). This generalizes a result of Babenko (1973, Proposition 4).

## 6. On the Uniqueness of *D*-solutions.

The aim of this section is to determine conditions under which a *D*-solution v is unique. The problem offers more or less the same type of technical difficulties encountered in the preceding section, as we are about to explain. Let us *denote by* $\mathscr{C}_v$ *the class of D-solutions achieving the same data as* v. We begin to give a quite general result (*cf.* Lemma 6.1) which ensures that if a *D*-solution satisfies some extra requirements at large distances (which a *priori* do *not* follow directly from Definition 4.1) and if the size of v in certain norms is suitably restricted (this condition is exactly analogous to that one imposes in the case Ω bounded) then, v is unique in an appropriate subclass $\mathscr{C}'_{v,q}$ of $\mathscr{C}_v$ (*cf.* Definition 6.1). Successively, we show that if the body force f satisfies certain summability restrictions at large distances and if f and $v_*-v_\infty$ are sufficiently small, then $\mathscr{C}'_{v,q}$ and $\mathscr{C}_v$ coincide for at least one value of q (=4); furthermore, any *D*-solution v corresponding to the above data satisfies the extra requirements so that v is unique in $\mathscr{C}_v$. However, under the stated conditions on f and $v_*$, *every D*-solution is in $\mathscr{C}_v$, so that we may conclude that v is unique in the class of all *D*-solutions corresponding to the given f, $v_*$ and $v_\infty$, *cf.* Theorem 6.1.

To accomplish our objective, however, we shall employ a method which is a bit different from that usually adopted for flows in bounded domains. This because the use of this latter method would lead to a uniqueness result which does not impose extra conditions directly on v but, rather, on v−w, $w\in\mathscr{C}_v$, in contrast to what stated previously. To see why this happens, we recall that the starting point of the "classical" method is the relation obtained by subtracting identity (4.1) written for w to the same identity written for v; such a relation reads

$$(\nabla u,\nabla\varphi)+(u\cdot\nabla u,\varphi)+(u\cdot\nabla v,\varphi)+(v\cdot\nabla u,\varphi)=0.$$

The next step is to substitute u for φ into this relation and this can be done via the usual approximating procedure which employs the continuity of the trilinear form (5.5). According to Lemma 5.1, we must then require some extra conditions on u. However, u is the *difference* of two generalized solutions and the method would lead to a uniqueness result different from that stated at the beginning of the current section.

The method we shall presently adopt is due to Galdi (1992b) and relies upon an idea introduced by Leray (1934, §32) in a completely different context, namely, that of local regularity of weak solutions to the *initial* value problem for the Navier–Stokes equations, and successively generalized by Serrin and Sather, *cf.* Serrin (1963, Theorem 6), Sather (1963, Theorem 5.1). In the case of steady flows in unbounded domains, the method has been also considered by Kozono & Sohr (1992) and Borchers, Galdi & Pileckas (1992).

In order to prove the main results, we begin to define a suitable subclass $\mathfrak{E}'_{v,q}$ of the class $\mathfrak{E}_v$.

**Definition 6.1.** $\mathfrak{E}'_{v,q}$ denotes the subclass of $\mathfrak{E}_v$ constitued by those *D*-solutions $\mathbf{w}$ such that:

(*i*) $\mathbf{w}$ satisfies the *generalized energy inequality*

$$|\mathbf{w}|_{1,2}+[\mathbf{f},\mathbf{w}-\mathbf{A}] \le (\nabla\mathbf{w},\nabla\mathbf{A}) -(\mathbf{w}\cdot\nabla\mathbf{A},\mathbf{w}-\mathbf{A}) \tag{6.1}$$

with $\mathbf{A}$ *an extension of* $\mathbf{v}_*$ and $\mathbf{v}_\infty$ *in the sense of* Definition 5.2.

(*ii*) the following properties hold

$$\begin{aligned} &\mathbf{w}-\mathbf{v}_\infty \in L^q(\Omega) \\ &\mathbf{v}_\infty\cdot\nabla\mathbf{w} \in L^{q'}(\Omega) \end{aligned} \tag{6.2}$$

for some $q>3/2$, $q'=q/(q-1)$.

**Remark 3.1.** The condition $q>3/2$ is required only because, if the flux *m* of $\mathbf{v}_*$ through $\partial\Omega$ is non-zero, $\mathbf{A}-\mathbf{v}_\infty$ is not summable at large distances for $q<3/2$ so that the term

$$(\mathbf{w}\cdot\nabla\mathbf{A},\mathbf{w}-\mathbf{A})$$

can be meaningless. On the other hand, if $m=0$ we can take $\mathbf{A}-\mathbf{v}_\infty$ of bounded support in $\Omega$ (see (5.16) and Definition 5.2) so that the restriction $q>3/2$ can be dropped and we may take an arbitrary $q>1$.

We are now in a position to prove the following

**Lemma 6.1.** *Let* $\Omega$ *be a locally lipschitzian, exterior domain of* $\mathbb{R}^3$. *Assume* $\mathbf{v}$ *is a D-solution to the Navier-Stokes problem* (0.1), (0.2) *corresponding to data*

$$f \in D_0^{-1,2}(\Omega),\ v_* \in W^{1/2}(\partial\Omega),\ v_\infty \in R^3,$$

*and such that*

$$v - v_\infty \in L^3(\Omega). \tag{6.3}$$

*Suppose, further,*

$$\begin{gathered} v - v_\infty \in L^q(\Omega) \\ v_\infty \cdot \nabla v \in L^{q'}(\Omega) \end{gathered} \tag{6.4}$$

*for some* $q > 3/2.$, $q' = q/(q-1)$. *Then, if*

$$\|v - v_\infty\|_3 < \frac{\sqrt{3}}{2}\,, \tag{6.5}$$

$v$ *is the unique solution in the class* $\mathcal{C}'_{v,q}$.

*Proof.* Let $w$ be any element from $\mathcal{C}'_{v,q}$. The field $w-A$ has zero trace at the boundary, is divergence free and vanishes at infinity in the sense of Definition 4.1 *(iv)*. Therefore, from the characterization (1.1), it follows

$$w - A \in \mathcal{D}_0^{1,2}(\Omega). \tag{6.6}$$

Furthermore, it is

$$w - A \in L^q(\Omega). \tag{6.7}$$

By (6.6), (6.7) and by Theorem 2.2 we may find a sequence $\{\varphi_k\} \subset \mathcal{D}(\Omega)$ converging to $w-A$ in $\mathcal{D}_0^{1,2}(\Omega) \cap L^q(\Omega)$. Replacing $\varphi_k$ for $\varphi$ into (1.2) and reasoning as in the proof of Theorem 5.1 (with slight difference in details) we deduce

$$(\nabla v, \nabla w) + [f, w-A] - (\nabla v, \nabla A) + (v \cdot \nabla v, w-A) = 0. \tag{6.8}$$

Likewise, observing that $v-A$ possesses the same summability properties of $w-A$ and that, in addition, by (6.3)

$$v - A \in L^3(\Omega),$$

one shows with no difficulty

$$(\nabla w, \nabla v) + [f, v-A] - (\nabla w, \nabla A) + (w \cdot \nabla w, v-A) = 0. \tag{6.9}$$

Finally, noticing that, by Theorem 1.1,

$$v - v_\infty \in L^6(\Omega),$$

from (6.3) we have, by interpolation,

$$v - v_\infty \in L^4(\Omega),$$

and by Theorem 5.2 we conclude

$$|v|_{1,2}^2 + [f, v-A] = (\nabla v, \nabla A) - (v \cdot \nabla A, v-A). \tag{6.10}$$

Addition of (6.8) and (6.9) yields

$$-2(\nabla w,\nabla v)-(v\cdot\nabla v,w-A)-(w\cdot\nabla w,v-A)+(\nabla v,\nabla A)$$
$$+(\nabla w,\nabla A)-[f,v-A]-[f,w-A]=0\ . \tag{6.11}$$

Summing side by side (6.1), (6.10) and (6.11) we deduce

$$|u|^2_{1,2}\leq (w\cdot\nabla w,v-A)+(v\cdot\nabla v,w-A)-(w\cdot\nabla A,w-A)-(v\cdot\nabla A,v-A)$$

with $u=w-v$. Adding and subtracting to the right hand side of this relation the quantity

$$(v\cdot\nabla w,v-A)+(u\cdot\nabla v,v-A)$$

(notice that each term in the sum is well-defined) we obtain

$$|u|^2_{1,2}\leq(u\cdot\nabla u,\ v-v_\infty)+(u\cdot\nabla u,v_\infty-A)+(u\cdot\nabla v,\ v-A)$$
$$+(v\cdot\nabla w,v-A)-(v\cdot\nabla A,v-A)-(w\cdot\nabla A,w-A)+(v\cdot\nabla v,w-A).$$

Since

$$(u\cdot\nabla u,v_\infty-A)+(u\cdot\nabla v,v-A)=(u\cdot\nabla v,v-v_\infty)+(u\cdot\nabla w,v_\infty-A),$$

it follows

$$|u|^2_{1,2}\leq(u\cdot\nabla u,v-v_\infty)+(u\cdot\nabla v,\ v-v_\infty)+(u\cdot\nabla w,v_\infty-A)$$
$$+(v\cdot\nabla(w-A),v-A)-(w\cdot\nabla A,w-A)+(v\cdot\nabla v,w-A). \tag{6.12}$$

The following identities hold

*(i)* $(u\cdot\nabla v,v-v_\infty)=0$

*(ii)* $(v\cdot\nabla v,w-A)=-(v\cdot\nabla(w-A),v-v_\infty)$

*(iii)* $(w\cdot\nabla A,w-A)=-(w\cdot\nabla(w-A),A-v_\infty)$

*(iv)* $(u\cdot\nabla w,v_\infty-A)=(u\cdot\nabla(w-A),v_\infty-A)$.

To show *(i)*, we let $\{u_k\}\subset\mathcal{D}(\Omega)$ be a sequence approximating u in $\mathcal{D}_0^{1,2}(\Omega)$. Set

$$\Omega_k=supp(u_k)$$

and denote by $\{v_m\}\subset C_0^\infty(\bar\Omega_k)$ a sequence converging –for each fixed k– to $v-v_\infty$ in $W^{1,2}(\bar\Omega_k)$. Clearly, for all $m\in\mathbb{N}$ we have

$$(u_k\cdot\nabla v_m,v_m)=\tfrac{1}{2}(u_k,\nabla v_m^2)=0$$

and so, passing to the limit $m\to\infty$, by Lemma 5.1 it is

$$(u_k\cdot\nabla v, v-v_\infty) = (u_k\cdot\nabla(v-v_\infty), v-v_\infty) = 0, \text{ for all } k\in\mathbb{N}.$$

This relation, along with Lemma 5.1 implies *(i)*. In a similar way, one proves *(ii)* and *(iii)*. Finally, consider the identity

$$(u\cdot\nabla w, v_\infty-A) = (u\cdot\nabla(w-A), v_\infty-A) - (u\cdot\nabla(b-v_\infty), A-v_\infty).$$

By a reasoning completely analogous to that used before one shows

$$(u\cdot\nabla(b-v_\infty), A-v_\infty) = 0,$$

so that also *(iv)* follows. Replacing *(i)*–*(iv)* into (6.12) we obtain

$$\begin{aligned}|u|^2_{1,2} &\le (u\cdot\nabla u, v-v_\infty) + (u\cdot\nabla(w-A), v_\infty-A)\\ &+(v\cdot\nabla(w-A), v_\infty-A) - (w\cdot\nabla(w-A), v_\infty-A)\\ &= (u\cdot\nabla u, v-v_\infty).\end{aligned} \tag{6.13}$$

From Lemma 5.1 it follows

$$|(u\cdot\nabla u, v-v_\infty)| \le \frac{2}{\sqrt{3}}|u|^2_{1,2}\|v-v_\infty\|_3,$$

which, once substituted into (6.13), furnishes

$$|u|^2_{1,2}(1-2(3)^{-1/2}\|v-v_\infty\|_3) \le 0. \tag{6.14}$$

Thus, if (6.5) holds, (6.14) implies $u(x)=0$ for a.a. $x\in\Omega$ and the theorem is completely proved.

The next theorem represents the main result of this section and shows that if $f$ has suitable summability properties at large distances and if, in addition, $f$ and $v_*-\lambda e_1$ (with $\lambda e_1 = v_\infty$) are sufficiently "small", the associated *D*-solution is unique in the class $\mathfrak{C}_v$ of *all* possible *D*-solutions corresponding to $f$, $v_*$ and $\lambda e_1$.

**Theorem 6.1** – *Let the assumptions of* Theorem 5.1 *be satisfied and assume, further that the data* $f$, $v_*$ *and* $v_\infty=\lambda e_1$ *verify*[1]

$$\begin{gathered} f\in L^{6/5}(\Omega)\\ \|f\|_{6/5}+\|v_*-\lambda e_1\|_{7/6,6/5,(\partial\Omega)} < \min\left\{\frac{a_1}{4c}, \frac{\sqrt{3}}{2}\right\}\end{gathered} \tag{6.15}$$

*where, for all* $\lambda\in(0,B]$, *it is* $c=c(\Omega,B)$ *and* $a_1=\min\{1,\sqrt{\lambda}\}$. *Then,* $v$ *is the only D-solution corresponding to the above data.*

---

[1] Observe that if $f\in L^{6/5}(\Omega)$ then $f\in D^{-1,2}(\Omega)$. Actually, by Sobolev inequality (1.4), for all $v\in D_0^{1,2}(\Omega)$ it follows

$$|[f,v]| \le \|f\|_{6/5}\|v\|_6 \le (2/\sqrt{3})\|f\|_{6/5}|v|_{1,2}.$$

*Proof.* We begin to show that, under the hypothesis (6.15) there exists a $D$-solution $\mathbf{w}$, say, verifying the condition

$$\|\mathbf{w}-\mathbf{v}_\infty\|_3 < \frac{\sqrt{3}}{2\mathcal{R}e}. \tag{6.16}$$

To this end, we employ the method of successive approximations along with Theorem 3.1 on the estimates for the Oseen problem. We introduce a sequence of approximating solutions $\{\mathbf{w}_m,\pi_m\}$, defined by $\mathbf{w}_0\equiv\pi_0\equiv 0$ and, for $m\geq 1$,

$$\begin{aligned} &\Delta\mathbf{w}_m-\lambda\frac{\partial\mathbf{w}_m}{\partial x_1} = \mathbf{w}_{m-1}\cdot\nabla\mathbf{w}_{m-1}+\nabla\pi_m-\mathbf{f},\\ &\nabla\cdot\mathbf{w}_m = 0,\\ &\mathbf{w}_m = \mathbf{v}_*-\mathbf{v}_\infty, \quad \text{at } \partial\Omega,\\ &\lim_{|x|\to\infty}\mathbf{w}_m(x) = 0. \end{aligned} \tag{6.17}$$

By Theorem 3.1 we know that, for m=1, there exists a solution $\mathbf{w}_1$, $p_1$ such that

$$\mathbf{w}_1\in L^3(\Omega)\cap D^{1,12/7}(\Omega)\cap D^{2,6/5}(\Omega),$$

$$\pi_1\in D^{1,6/5}(\Omega),$$

and obeying the estimate

$$\begin{aligned} &a_1\|\mathbf{w}_1\|_3+a_2|\mathbf{w}_1|_{1,12/7}+|\mathbf{w}|_{2,6/5}+|\pi_1|_{1,6/5}\\ &\leq c_1(\|\mathbf{f}\|_{6/5}+\|\mathbf{v}_*-\mathbf{v}_\infty\|_{7/6,6/5,(\partial\Omega)} \equiv c_1 D \end{aligned} \tag{6.18}$$

where $c_1=c_1(q,\Omega)$ is independent of $\lambda$ for $\lambda\in(0,B]$ and

$$a_1=\min\{1,\lambda^{1/2}\}$$
$$a_2=\min\{1,\lambda^{1/4}\}.$$

Since, by Theorem 1.1,

$$|\mathbf{w}_1|_{1,2}\leq c_1|\mathbf{w}_1|_{2,6/5}, \tag{6.19}$$

(6.18) furnishes, in particular,

$$a_1\|\mathbf{w}_1\|_3+|\mathbf{w}_1|_{1,2}+|\mathbf{w}_1|_{2,6/5}+|\pi_1|_{1,6/5}\leq cD. \tag{6.20}$$

We next show, by induction, that the following inequality is verified, for all $m\in\mathbb{N}$

$$a_1\|\mathbf{w}_m\|_3+|\mathbf{w}_m|_{1,2}+|\mathbf{w}_m|_{2,6/5}+|\pi_m|_{1,6/5}\leq 2cD, \tag{6.21}$$

provided $D$ is "small enough". Thus, assuming $\mathbf{w}_m,\pi_m$ obey (6.21), by Theorem 3.1 and (6.19) we recover

$$a_1\|w_{m+1}\|_3+|w_{m+1}|_{1,2}+|w_{m+1}|_{2,6/5}+|\pi_{m+1}|_{1,6/5}\le c(D+\|w_m\cdot\nabla w_m\|_{6/5}). \qquad (6.22)$$

Now, we have

$$\|w_m\cdot\nabla w_m\|_{6/5}\le\|w_m\|_3|w_m|_{1,2}$$

and, by induction hypothesis,

$$\|w_m\cdot\nabla w_m\|_{6/5}\le 4c^2D^2/a_1,$$

so that (6.22) shows that, if

$$D< a_1/4c\mathcal{R}e, \qquad (6.23)$$

inequality (6.21) is satisfied for all $m\in\mathbb{N}$. It is easy now to prove that $\{w_m,\pi_m\}$ is a Cauchy sequence in the space

$$\mathcal{S}\equiv L^3(\Omega)\cap D^{1,2}(\Omega)\cap D^{2,6/5}(\Omega)\times D^{1,6/5}(\Omega).$$

In fact, from (6.17) we deduce

$$a_1\|w_{m+1}-w_m\|_3+|w_{m+1}-w_m|_{1,2}+|w_{m+1}-w_m|_{2,6/5}+|\pi_{m+1}-\pi_m|_{1,6/5}$$
$$\le c\|w_m.\nabla w_m-w_{m-1}\cdot\nabla w_{m-1}\|_{6/5},$$

and since

$$\|w_m\cdot\nabla w_m-w_{m-1}\cdot\nabla w_{m-1}\|_{6/5}\le\|w_m-w_{m-1}\|_3|w_m|_{1,2}+\|w_m\|_3|w_m-w_{m-1}|_{1,2},$$

in view of (6.21) we conclude, for all $m\ge1$,

$$a_1\|w_{m+1}-w_m\|_3+|w_{m+1}-w_m|_{1,2}+|w_{m+1}-w_m|_{2,6/5}+|\pi_{m+1}-\pi_m|_{1,6/5}$$
$$\le(2CD/a_1)(a_1\|w_m-w_{m-1}\|_3+|w_m-w_{m-1}|_{1,2}).$$

From this inequality we receive, for all $m\ge1$,

$$a_1\|w_{m+1}-w_m\|_3+|w_{m+1}-w_m|_{1,2}+|w_{m+1}-w_m|_{2,6/5}+|\pi_{m+1}-\pi_m|_{1,6/5}\le(2CD/a_1)^{m+1}$$

which, by (6.23) and by virtue of a standard argument, implies that $\{w_m,\pi_m\}$ ios Cauchy in the space $\mathcal{S}$. Denoting by $w,\pi$ the limiting field, we then have

$$w\in L^3(\Omega)\cap D^{1,2}(\Omega)\cap D^{2,6/5}(\Omega)$$
$$\pi\in D^{1,6/5}(\Omega)$$

and, moreover, $w$, $\pi$ obey the estimate (6.21). In addition, from the same argument employed in the proof of Theorem 5.1 we find

$$w-v_\infty\in L^4(\Omega)$$
$$v_\infty\cdot\nabla w\in L^{4/3}(\Omega) \qquad (6.24)$$

and, if the data satisfy $(6.15)_2$, we have also

$$\|w-v_\infty\|_3 < \frac{\sqrt{3}}{2\mathcal{R}e} \tag{6.25}$$

Let now $v$ denote *any* $D$-solution corresponding to $f, v_*$ and $v_\infty$. From Remark 5.6 we find

$$\begin{aligned} &v-v_\infty \in L^4(\Omega) \\ &v_\infty \cdot \nabla v \in L^{4/3}(\Omega) \end{aligned} \tag{6.26}$$

and, consequently, we have

$$\mathfrak{E}_v = \mathfrak{E}'_{v,4}. \tag{6.27}$$

Thus, by Lemma 6.1 we at once obtain $w=v$ which furnishes, in particular,

$$\|v-v_\infty\|_3 < \frac{\sqrt{3}}{2\mathcal{R}e}\,. \tag{6.28}$$

The theorem becomes then a consequence of (6.27), (6.28) and of Lemma 6.1.

## References

BABENKO, K.I., 1973, On Stationary Solutions of the Problem of Flow Past a Body of a Viscous Incompressible Fluid, *Mat. Sbornik,* 91 (133), p.1. English Translation in *Math.* USSR *Sbornik* 20 (1973), p.1.

BOGOVSKIĬ, M.E., 1979, Solution of the First Boundary Value Problem for the Equation of Continuity of an Incompressible Medium, *Dokl. Akad Nauk* SSSR, 248, (5), p.817. English translation in *Soviet Math Dokl.*, 20 (1979), p.1094.

BOGOVSKIĬ, M.E., 1980, Solution of Some Vector Analysis Problems Connected with Operators Div and Grad, *Trudy Seminar S.L.Sobolev,* #1, 80, *Akademia Nauk SSSR, Sibirskoe Otdelnie Matematiki, Nowosibirsk,* p.5 (in Russian).

BORCHERS, W., GALDI, G.P. and PILECKAS, K., 1992, On the Uniqueness of Leray-Hopf Solutions for the Flow Through an Aperture, *Arch. Ratl Mech. Anal.*, in the Press

CHANG, I.D. and FINN, R., 1961, On the Solutions of a Class of Equations Occurring in Continuum Mechanics with Application to the Stokes Paradox, *Arch. Ratl Mech. Anal.*, 7, p.388.

ERIG, W., 1982, Die Gleichungen von Stokes un die Bogovskiĭ-Formel, *Diplomarbeit, Universität Paderborn.*

FINN, R., 1965, On the Exterior Stationary Problem for the Navier-Stokes Equations, and Associated Perturbation Problems, *Arch. Ratl Mech. Anal.* 19, p.363.

FUJITA, H., 1961, On the Existence and Regularity of the Steady-State Solutions of the Navier-Stokes Equations, *J.Fac.Sci.Univ.Tokyo,* Sect. I, 9, p.59.

GALDI, G.P., 1992a, On the Oseen Boundary-Value Problem in Exterior Domains, *Proc. of the Oberwolfach Meeting "The Navier-Stokes Equations: Theory and Numerical Methods", J.G.Heywood, K.Masuda, R.Rautmann and V.A.Solonnikov Eds., Springer-Verlag Lecture Notes in Mathematics,* in the Press.

GALDI, G.P., 1992b, On the Asymptotic Properties of Leray's Solutions to the Exterior Steady Three-dimensional Navier-Stokes equations with Zero Velocity at Infinity, *Proc. Conference "Degenerate Diffusions", W.Ni, L.A.Peletier and J.L.Vasquez Eds, Springer - Verlag,* in the Press.

GALDI, G.P. (*forthcoming*) An Introduction to the Mathematical Theory of the Navier-Stokes Equations, *Springer Tracts in Natural Phylosophy.*

GALDI, G.P. and Maremonti, P., 1986, Monotonic Decreasing and Asymptotic Behavior of the Kinetic Energy for Weak Solutions of the Navier-Stokes Equations in Exterior Domains, *Arch. Ratl Mech. Anal.*, 94, p.253

KOZONO, H. and SOHR, H., 1992, On Stationary Navier Stokes Equations in Unbounded Domains, *Ricerche Mat.*, in the Press.

HOPF, E., 1951, Über die Anfangswertaufgabe für die Hydrodynamischen Grundleichungen, *Math. Nachr.*, 4, p.213.

LERAY, J., 1933, Étude de Diverses Équations Intégrales non Linéaires et de Quelques Problèmes que Pose l'Hydrodynamique, *J. Math. Pures Appl.*, 12, p.1.

LERAY, J., 1934, Sour le Mouvement d'une Liquide Visqueux Emplissant l'Espace, *Acta Math.* 63, p.193.

NEČAS, J., 1967, Les Méthodes Directes en Théorie des Équations Elliptiques, *Masson et C*$^{ie}$.

NIRENBERG, L., 1959, On Elliptic Partial Differential Equations, *Ann. Sc. Norm. Pisa*, 13, (3), p.115.

SATHER, J., 1964, The Initial Boundary Value Problem for the Navier-Stokes Equations in Regions with Moving Boundaries, *Ph.D. Thesis, University of Minnesota.*

SERRIN, J., 1963, The Initial Value Problem for the Navier-Stokes Equations, *Nonlinear Problems, R.E. Langer Ed., University of Wisconsin Press,* p.69.

SOBOLEV, S., 1963, Denseness of Finite Fields in the Space $L_p^m(E_n)$, *Sib. Mat. Zh.*, 4, p.673 (in Russian).

SOLONNIKOV, V.A. and ŠČADILOV, V.E., 1973, On a Boundary Value Problem for Stationary Navier-Stokes Equations, *Tr. Math. Inst. Steklov* 125, p. 176. English Translation in *Proc. Steklov Inst. of Math.* 125 (1973), p. 186.

**STEIN, E., 1970, Singular Integrals and Differentiability properties of Functions, *Princeton University Press.***

**TEMAM, R., 1977, Navier-Stokes Equations, *North Holland.***

# On the Asymptotic Structure of *D*-Solutions to Steady Navier-Stokes Equations in Exterior Domains

Giovanni P. Galdi
Istituto di Ingegneria
Via Scandiana 21
Universitá degli Studi
Ferrara, Italia

## Introduction.

In this paper we continue and, to some extent, conclude the research on *D*-solutions initiated in the preceding article of Galdi, hereafter denoted by [G]. Specifically, we shall prove that every *D*-solution $\mathbf{v}$ to the (steady) Navier-Stokes problem in a three dimensional, exterior domain $\Omega$, *cf.* [G, Definition 4.1], corresponding to a non-zero velocity $\mathbf{v}_\infty$ at infinity and to sufficiently smooth data with the body force of bounded support, presents the same asymptotic structure of the fundamental solution of the associated linearized problem, *i.e.*, as $|\mathbf{x}|\to\infty$,

$$\mathbf{v}(\mathbf{x})=\mathbf{v}_\infty+\mathbf{m}\cdot\mathbf{E}(\mathbf{x})+O(|\mathbf{x}|^{-3/2+\delta}) \tag{0.1}$$

where $\mathbf{m}$ is a constant vector, $\mathbf{E}$ is the Oseen fundamental tensor and $\delta$ is an arbitrary positive number.

Such a result, obtained by an independent (and more intricated) method by Babenko (1973), is here proved by means of the estimates for the Oseen system recently established by Galdi (1992). The main difficulty in the present paper -as in Babenko's- is to show that, in the exterior of a ball of sufficiently large radius, $\mathbf{v}-\mathbf{v}_\infty$ belongs to the Lebesgue space $L^q$ for *some* q *strictly* less than 4 [1].The proof of this property is achieved by suitably adapting the estimates of Galdi (1992) to a certain linearized version of the Navier-Stokes system.

---

[1] The demonstration of such a statement is the content of pp.11-21 of Babenko's paper. However, the present author regrets that some steps in the proof are obscure to him.

The plan of the paper is as follows: In Section 1 we recall some classical results concerning weighted inequalities, the Oseen fundamental tensor and the representation of *D*-solutions. We also furnish (Lemma 1.1) a suitable approximation of certain solenoidal functions. Successively, in Section 2, we establish some preliminary results concerning the pointwise behaviour of a *D*-solution and of the associated pressure field p, under very general assumptions on the body force. In Section 4 we prove the fundamental result. Precisely, with the help of Lemma 1.1 and of the results of Galdi (1992) we show that, if the data satisfy certain "natural" conditions, any corresponding *D*-solution enjoys the same summability properties of the Oseen fundamental tensor. With such a result in hands, it is then quite simple to obtain the asymptotic structure (0.1) for a *D*-solution corresponding to external forces of bounded support, and this is the content of Section 4.

## 1. Preparatory Results.

In the present section we shall collect some basic results which will be needed in the sequel. All of them are well known, except for that proved in Lemma 1.1. The notations we will adopt are the same ones introduced in [G, Section 1].

We begin with a weighted inequality whose proof can be deduced from the works of Finn (1965, Corollary 2.2a) and of Galdi & Maremonti (1986, Lemma 1.3), *cf.* Galdi (forthcoming). For a>0 we set

$$B_a(x) = \{x\in R^3 : |x-y|<a\}, \quad B_a\equiv B_a(0).$$

**Theorem 1.4.** *Let $\Omega$ be a three dimensional, exterior domain. Assume*

$$u\in D^{1,2}(\Omega).$$

*Then, there exists $u_0\in R$ such that, setting $w=u-u_0$, the following properties are true. For any $x_0\in R^3$, we have*

$$w|x-x_0|^{-1}\in L^2(\Omega^R(x_0))$$

*where*

$$\Omega^R(x_0) \equiv \Omega - B_R(x_0)\ ,$$
$$B_R(x_0) \supset \Omega^c,$$

*and, moreover,*

$$\left(\int_{\Omega^R(x_0)} |w(x)/(x-x_0)|^2 dx\right)^{1/2} \le q/(n-q)|w|_{1,2,\Omega^R(x_0)}. \tag{1.1}$$

*Furthermore, if* $|x_0|=\alpha R$, *for some* $\alpha\ge\alpha_0>1$ *and some* $R>\delta(\Omega^c)$, *we have*

$$\left(\int_{\Omega^R} |w(x)/(x-x_0)|^2 dx\right)^{1/2} \le c|w|_{1,2,\Omega^R} \tag{1.2}$$

*where* $c=c(\alpha_0)$.

Our next objective is to recall some classical representation formulas of solutions to the Navier-Stokes equations in an exterior domain. They are due to Finn (1959), even though the assumptions under which we state them here are slightly more general than those of Finn, *cf.* Galdi (forthcoming). To this end, we recall the definition of *Oseen fundamental solution, cf.* Oseen (1927, §4). As is well known, by this we mean a second order tensor field $E=E(x-y)$ and a vector field $e=e(x-y)$ satisfying the following system

$$\begin{aligned} (\Delta_y+\lambda\frac{\partial}{\partial y_1})E_{ij}(x-y) -\frac{\partial}{\partial y_i} e_j(x-y) &= \delta_{ij}\delta(x-y) \\ \frac{\partial}{\partial y_\ell} E_{\ell j}(x-y) &= 0 \end{aligned} \qquad x,y\in R^3, \tag{1.3}$$

where $\delta_{ij}$ is the unit matrix and $\delta(\xi)$ is the Dirac function. A solution to (1.3) is explicitly found and one has

$$\begin{aligned} E_{ij}(x-y) &= (\delta_{ij}\Delta_y - \frac{\partial^2}{\partial y_i \partial y_j})\Phi(x-y) \\ e_j(x-y) &= \frac{1}{4\pi}\frac{x_j-y_j}{|x-y|^3} \end{aligned} \tag{1.4}$$

where

$$\Phi(x-y)=-\frac{1}{4\pi\lambda}\int_0^{\lambda(|x-y|-(x_1-y_1))/2} \frac{1-e^{-\tau}}{\tau}\,d\tau. \tag{1.4$'$}$$

We reproduce some elementary estimates of the tensor E, *cf.* Finn (1965), Galdi (forthcoming). Specifically, it is

$$
\begin{aligned}
&|E(x-y)| \le c|x-y|^{-1}, \\
&|\nabla E(x-y)| \le c|x-y|^{-3/2} \qquad (1.5)\\
&\int_{\partial B_R(x)} |\nabla E(x-y)|\, d\sigma_y \le cR^{-1/2}.
\end{aligned}
$$

with c a positive constant independent of x,y. Moreover, for $\mathcal{A}$ the exterior of ball of sufficiently large radius, we have

$$
\begin{aligned}
&E(y) \in L^r(\mathcal{A}), \quad \text{for all } r>2. \\
&\frac{\partial E(y)}{\partial y_1} \in L^s(\mathcal{A}), \quad \text{for all } s>1, \qquad (1.5)'\\
&\frac{\partial E(y)}{\partial y_i} \in L^t(\mathcal{A}), \quad \text{for all } t>4/3,\ i=2,3, \\
&e(y) \in L^\sigma(\mathcal{A}), \quad \text{for all } \sigma>3/2.
\end{aligned}
$$

The following representation formula holds.

**Theorem 1.2.** *Let* $v$ *be a D-solution to the Navier-Stokes equations in a domain* $\Omega$ *of class* $C^2$, *with*

$$v \in W^{2,r}(\Omega_R),\ \textit{for some } R>\delta(\Omega^c) \textit{ and } r\in(1,\infty).$$

*Then, if* $f$ *is of bounded support and*

$$f \in L^s(\Omega),\ \textit{for some } s\in(1,\infty),$$

*setting* $u=v-v_\infty$, *the following asymptotic representations hold as* $|x|\to\infty$

$$v_j(x) = v_{\infty j} + M_i E_{ij}(x) + \int_\Omega E_{ij}(x-y)u(y)\cdot\nabla u_i(y)dy + \sigma_j(x) \qquad (1.6)$$

$$v_j(x) = v_{\infty j} + m_i E_{ij}(x) - \int_\Omega u_i(y)u(y)\cdot\nabla_y E_{ij}(x-y)dy + \eta_j(x) \qquad (1.7)$$

*where*

$$
\begin{aligned}
M &= -\int_{\partial\Omega} [T(v,p)\cdot n - \lambda v v_\infty\cdot n] + \int_\Omega f \\
& \qquad\qquad\qquad\qquad\qquad\qquad\qquad\qquad (1.8)\\
m &= -\int_{\partial\Omega} [T(v,p)\cdot n - \lambda(v-v_\infty)\cdot n] + \int_\Omega f
\end{aligned}
$$

*and*

$$\sigma(x),\ \eta(x) = O(|x|^{-3/2}). \qquad (1.9)$$

The next (and final) result of the present section concerns a suitable approximation of certain solenoidal vector fields.

**Lemma 1.1.** *Let $\Omega$ be a locally lipschitian, exterior domain of $\mathbb{R}^3$ and let $u\in L^\infty(\Omega_R)\cap L^4(\Omega)$, for all $R>\delta(\Omega^c)$, with $\nabla\cdot u=0$[2] in $\Omega$. Then, there exists a sequence $\{u_k\}$ of solenoidal functions in $\Omega$ such that*

$$\begin{aligned} &u_k\in L^3(\Omega), \qquad \textit{for all } k\in\mathbb{N};\\ &\lim_{k\to\infty} \|u_k-u\|_4 = 0. \end{aligned} \tag{1.10}$$

*Furthermore, given $\varepsilon>0$ the sequence $\{u_k\}$ can be chosen such as to satisfy the following properties, for all $k\in\mathbb{N}$,*

$$\begin{aligned} &u_k = u_k^{(1)}+u_k^{(2)},\\ &\|u_k^{(1)}\|_4 < \varepsilon,\\ &u_k^{(2)}\in L^\infty(\Omega_\rho),\\ &\mathrm{supp}(u_k^{(2)})\subseteq \overline{\Omega}_\rho\\ &\|u_k^{(2)}\|_{\infty,\Omega_\rho} \le \|u\|_{\infty,\Omega_\rho} \end{aligned} \tag{1.11}$$

*where $\rho$ depends only on $\varepsilon$ and $u$.*

*Proof.* Let $\varphi\in C^1(\mathbb{R})$ be a non-negative function with $\varphi(t)=1$ for $t\le 1/2$ and $\varphi(t)=0$ for $t\ge 1$. For $a>0$ we set

$$\varphi_a(x) = \varphi\left(\frac{|x|}{a}\right), \qquad x\in\mathbb{R}^3.$$

We choose $\rho=\rho(\varepsilon,u)$ such that

$$\|(1-\varphi_\rho)u\|_4 < \varepsilon/2. \tag{1.12}$$

Let $\{R_k\}$ be an increasing, unbounded sequence of positive numbers with $R_1>2\rho$ and let $z_k$ denote a solution to the problem

$$\begin{aligned} &\nabla\cdot z_k = -\nabla\cdot(\varphi_{R_k} u),\\ &z_k\in D_0^{1,3/2}(\Omega)\cap D_0^{1,12/7}(\Omega),\\ &|z_k|_{1,3/2} \le c\|\nabla\cdot(\varphi_{R_k} u)\|_{3/2},\\ &|z_k|_{1,12/7} \le c\|\nabla\cdot(\varphi_{R_k} u)\|_{12/7}. \end{aligned} \tag{1.13}$$

Since $u$ is solenoidal, it is

$$\nabla\cdot(\varphi_{R_k} u) = u\cdot\nabla\varphi_{R_k},$$

[2] In the sense of distribution.

and, since the support of $\nabla\varphi_{R_k}$ is contained in $\Omega_{R_k,2R_k}$, by the assumptions made on u we have

$$\nabla\cdot(\varphi_{R_k} u)\in L^{3/2}(\Omega)\cap L^{12/7}(\Omega).$$

Thus, in view of the results of Bogovskiĭ (1980, Theorem 4) there exists a sequence of fields $z_k$ satisfying (1.13). From $(1.13)_{3,4}$ and Sobolev inequality (1.4) of [G] it follows

$$\begin{aligned} &\|z_k\|_4 \le c\gamma\|u.\nabla\varphi_{R_k}\|_{12/7} \\ &z_k\in L^3(\Omega), \text{ for all } k\in\mathbb{N}. \end{aligned} \tag{1.14}$$

Also, by Hölder inequality, it is

$$\|u\cdot\nabla\varphi_{R_k}\|_{12/7} \le \|\nabla\varphi_{R_k}\|_{3,\Omega_{R_k,2R_k}} \|u\|_{4,\Omega_{R_k,2R_k}} \tag{1.15}$$

and so, noticing that

$$|\nabla\varphi_{R_k}| \le c_1/R_k$$

for a constant $c_1$ independent of k, from (1.14), (1.15) we deduce for all $k\in\mathbb{N}$

$$\|z_k\|_4 \le c_2\|u\|_{4,\Omega_{R_k,2R_k}} \tag{1.16}$$

Define

$$u_k = \varphi_{R_k} u+z_k \tag{1.17}$$

From $(1.14)_2$ and (1.16) we at once recover (1.10). Furthermore, setting

$$\begin{aligned} &u_k^{(1)}= \varphi_{R_k}(1-\varphi_\rho)u+z_k \\ &u_k^{(2)}= \varphi_\rho u, \end{aligned} \tag{1.18}$$

and noticing that $\varphi_{R_k}\varphi_\rho=\varphi_\rho$ for all $k\in\mathbb{N}$, we obtain $(1.11)_{1,3,4}$.

From (1.18), (1.12) and (1.16) it follows

$$\|u_k^{(1)}\|_4 \le \|(1-\varphi_\rho)u\|_4+\|z_k\|_4 < (\varepsilon/2)+c_2\|u\|_{4,\Omega_{R_k,2R_k}} \tag{1.19}$$

and so, choosing $k\ge\bar{k}(\varepsilon)\in\mathbb{N}$ such that

$$c_2\|u\|_{4,\Omega_{R_k,2R_k}} < \frac{\varepsilon}{2},$$

from (1.19) we conclude

$$\|u_k^{(1)}\|_4 < \varepsilon.$$

Thus, relabelling the sequence $\{u_k\}$, we obtain a sequence satisfying all assertions in the lemma which is, therefore, completely proved.

## 2. Asymptotic Behaviour of *D*-solutions. Preliminary Results.

In the present section we commence the study of the asymptotic behaviour of a *D*-solution $v$ to the Navier-Stokes equations. Specifically, if the body force possesses a certain degree of summability at large distances, we begin to prove that $v$ and the corresponding pressure $p$, *cf.* [G, Remark 4.1] satisfy for some $p_0 \in \mathbb{R}$

$$v(x) \to v_\infty \ , \ \nabla v(x) \to 0, \ p(x) \to p_0 \quad \text{as } |x| \to \infty, \text{ uniformly.}$$

Successively, we show that if on $u \equiv v - v_\infty$ we assume some summability at infinity then $u$, $\nabla u$ and $p - p_0$ enjoy the same summability properties of the Oseen fundamental solution $\mathbf{E}, \mathbf{e}$, *cf.* (1.5)'.

**Lemma 2.1.** *Assume that there are* $R > \delta(\Omega^c)$ *and* $q \in (3/2, \infty)$ *such that*

$$f \in L^q(\Omega^R). \tag{2.1}$$

*Then, any D-solution corresponding to* $f$ *satisfies*

$$\lim_{|x| \to \infty} v(x) = v_\infty, \tag{2.2}$$

*uniformly. Moreover, if for some* $r \in (3, \infty)$ *and* $q \in (3/2, 2)$ *it is*

$$f \in L^r(\Omega^R) \cap f \in L^q(\Omega^R), \tag{2.3}$$

*we have also for some* $p_0 \in \mathbb{R}$

$$\lim_{|x| \to \infty} \nabla v(x) = 0, \qquad \lim_{|x| \to \infty} p(x) = p_0, \tag{2.4}$$

*uniformly.*

*Proof.* From the work of Fujita (1961, Lemma 4.2 and p.98) we know that for a.a. $x \in \Omega$ with $dist(x, \partial\Omega_R) > d$, the following representation formula holds

$$\begin{aligned} u_j(x) &= \int_{B_d(x)} E^{(d)}_{ij}(x-y)[f_i(y) + u(y) \cdot \nabla u_i(y)]dy - \int_{\beta(x)} H^{(d)}_{ij}(x-y) u_i(y) dy \\ &\equiv I_1(x) + I_2(x) + I_3(x). \end{aligned} \tag{2.5}$$

Here $u = v - v_\infty$, $B_a(x)$ is the ball of radius $a$ centered at $x$ while $\beta(x) = B_d(x)/B_{d/2}(x)$. Furthermore, $E^{(d)}_{ij}(x-y)$ -the truncated fundamental solution of Oseen equation- is a second order tensor, depending only on x-y, which vanishes for $|x-y|>d$ and, for sufficiently small d, satisfies the estimates

$$\left.\begin{array}{l} |E^{(d)}_{ij}(x-y)|\le c|x-y|^{-1} \\ |\nabla E^{(d)}_{ij}(x-y)|\le c|x-y|^{-2} \end{array}\right\}, \quad \text{for all } y\in B_d(x). \tag{2.6}$$

Finally, $H^{(d)}_{ij}(x-y)$ is an indefinitely differentiable function vanishing unless $d/2<|x-y|<d$ and satisfying

$$|H^{(d)}_{ij}(x-y)| + |\nabla H^{(d)}_{ij}(x-y)|\le M \tag{2.7}$$

with $M$ independent of x,y and with $\nabla$ operating either on x or on y. Using Hölder inequality and taking into account $(2.6)_1$, we readily deduce

$$|I_1(x)|\le c_1\|f\|_{q,B_d(x)} \tag{2.8}$$

and

$$|I_2(x)|\le c_2\|u/|x-y|\|_{2,B_d(x)}|u|_{1,2,B_d(x)}.$$

On the other hand, by Theorem 1.1, it is

$$\|u/|x-y|\|_{2,B_d(x)}\le c_4|u|_{1,2},$$

and therefore

$$|I_2(x)|\le c_5|u|_{1,2,B_d(x)}. \tag{2.9}$$

Finally, from (2.7) and Hölder inequality we find

$$|I_3(x)|\le c_6\|u\|_{6,B_d(x)}, \tag{2.10}$$

and so, recalling that by [G, Theorem 1.1] it is

$$u\in L^6(\Omega^R) \tag{2.11}$$

property (2.2) follows from (2.1), (2.5) and (2.8)-(2.11). Let us now show the first relation in (2.4). We begin to notice that, by (2.3) and by [G, Theorem 4.2] we have for any fixed $R_1>R$ and some $q\in(3/2,2]$

$$v\in W^{2,q}(\Omega_{R_1,R_2}),\ p\in W^{1,q}(\Omega_{R_1,R_2}), \text{ for all } R_2>R_1. \tag{2.12}$$

By the Sobolev imbedding theorem, these properties imply, in particular

$$v\in L^\infty(\Omega_{R_1,R_2}), \text{ for all } R_2>R_1$$

and so, because of (2.1),

$$u\in L^\infty(\Omega^{R_1}). \tag{2.13}$$

Consequently, from the inequality

$$\|u\cdot\nabla u\|_s \le \|u\|_{2s/(2-s)}|u|_{1,2},\quad 1\le s\le 2,$$

it follows

$$u\cdot\nabla u\in L^s(\Omega^R{}_1),\ \text{for all } s\in[3/2,2]. \tag{2.14}$$

From (2.3) and (2.13) we then recover

$$F\equiv f+u\cdot\nabla u\in L^q(\Omega^R{}_1),\ \text{for some } q\in(3/2,2)$$

and thus, recalling that u satisfies a.e. in $\Omega^R{}_1$ the Oseen problem (3.1) of [G], by (2.12) and by [G, Theorems 3.1, 3.2] it follows

$$v\in D^{2,q}(\Omega^R{}_1),\ p\in D^{1,q}(\Omega^R{}_1) \tag{2.15}$$

With the help of [G, Theorem 1.1] we then obtain for some $p_0\in\mathbb{R}$

$$v\in D^{1,3q/(3-q)}(\Omega^R{}_1),\quad p-p_0\in L^{3q/(3-q)}(\Omega^R{}_1), \tag{2.16}$$

and we conclude, in particular,

$$u\cdot\nabla u\in L^{3q/(3-q)}(\Omega^R{}_1). \tag{2.17}$$

Now, from (2.5) we deduce (with $D_k=\partial/\partial x_k$)

$$D_k u_j(x)=\int_{B_d(x)} D_k E^{(d)}_{1j}(x-y)F_1(y)dy - \int_{\beta(x)} D_k H^{(d)}_{1j}(x-y)u_1(y)dy. \tag{2.18}$$

Using Hölder inequality and $(2.6)_2$ we have

$$\Big|\int_{B_d(x)} D_k E^{(d)}_{1j}(x-y)F_1(y)dy\Big| \le c_7 \big\||x-y|^{-2}\big\|_{r/(r-1),B_d(x)}\|f\|_{r,B_d(x)}$$
$$\big\||x-y|^{-2}\big\|_{3q/(4q-3),B_d(x)}\|u\cdot\nabla u\|_{3q/(3-q),B_d(x)}\Big) \tag{2.19}$$

and so, being r, 3q/(3–q)>3, by the assumptions on f and by (2.17) we obtain that the first term at the right hand side of (2.18) tends to zero uniformly, as $|x|\to\infty$. Since the second term tends to zero as well –as a consequence of (2.7) and of (2.11)– we conclude the validity of the first relation in (2.4). We now come to the pressure. From (2.12) and from the Navier-Stokes equations, we derive that $\pi=p-p_0$ is a solution (in the sense of distributions) to the following equation

$$\Delta\pi = \nabla\cdot(f+u\cdot\nabla u)\quad \text{in } \Omega^R{}_1.$$

Therefore, by a reasoning entirely analogous to that used by Fujita to derive (2.5) one can show, *cf*. Galdi (forthcoming) for details,

$$\pi(x)=-\int_{B_d(x)} D_1 \mathcal{E}^{(d)}(x-y)[f_1(y)+u(y)\cdot\nabla u_1(y)]dy-\int_{\beta(x)} h^{(d)}(x-y)\pi(y)dy,$$

where $\mathcal{E}^{(d)}$ –the truncated fundamental solution of Laplace's equation– and $h^{(d)}$ enjoy the same properties stated for $E_{ij}^{(d)}$ and $H_{ij}^{(d)}$, respectively. Therefore, using the summability property (2.16) and proceeding as in (2.19) we show also the second relation into (2.4) and the lemma is completely proved.

In the next lemma we show that, if $\mathbf{f}$ verifies somewhat more stringent assumptions than those of Lemma 2.1 and $\mathbf{v}_*$ has a mild degree of regularity, every corresponding *D*-solution $\mathbf{v}$ in a $C^2$-smooth domain $\Omega$ with $\mathbf{v}-\mathbf{v}_\infty \in L^3(\Omega)$ satisfies the same summability properties (1.5)' of the fundamental tensor $\mathbf{E}$.

**Lemma 2.2.** *Let $\Omega$ be a $C^2$-smooth, exterior domain and assume*

$$\begin{aligned} &\mathbf{f}\in L^q(\Omega),\\ &\mathbf{v}_*\in W^{2-1/q_0,1/q_0}(\partial\Omega), \end{aligned} \tag{2.20}$$

*for some* $q_0>3$, *and all* $q\in(1,q_0]$. *Then every D-solution* $\mathbf{v}$ *to the Navier-Stokes problem corresponding to* $\mathbf{f}$, $\mathbf{v}_*$ *and* $\mathbf{v}_\infty$ *and such that*

$$\mathbf{v}-\mathbf{v}_\infty\in L^3(\Omega)$$

*satisfies*

$$\begin{aligned} &\mathbf{v}-\mathbf{v}_\infty\in L^r(\Omega),\\ &\frac{\partial \mathbf{v}}{\partial x_i}\in L^s(\Omega),\\ &\mathbf{v}\in D^{1,t}(\Omega),\\ &p\in L^\sigma(\Omega), \end{aligned} \tag{2.21}$$

*for all* $r\in(2,\infty]$, $s\in(1,\infty]$, $t\in(4/3,\infty]$, *where* p *is (up to a constant) the pressure field associated to* $\mathbf{v}$.

*Proof.* From [G, Theorem 4.2] we know that, under the stated assumptions on the data and on $\Omega$, it is

$$\mathbf{v}\in W^{2,q_0}(\Omega_R),\ p\in W^{1,q_0}(\Omega_R),\ \text{for all } R>\delta(\Omega^c).$$

These conditions along with the Sobolev imbedding theorem imply

$$\mathbf{v},\ \nabla\mathbf{v},\ p\in L^\infty(\Omega_R)\ \text{for all } R>\delta(\Omega^c). \tag{2.22}$$

We next observe that, setting, as usual, $\mathbf{u}=\mathbf{v}-\mathbf{v}_\infty$, in view of Hölder inequality and by hypothesis, it is

$$\|\mathbf{u}\cdot\nabla\mathbf{u}\|_{6/5}\le\|\mathbf{u}\|_3\,|\mathbf{u}|_{1,2} \tag{2.23}$$

so that

$$\mathbf{F}=\mathbf{f}+\mathbf{u}\cdot\nabla\mathbf{u}\in L^{6/5}(\Omega). \tag{2.24}$$

Now, as already observed in other circumstances, $\mathbf{u}$ solves the following Oseen problem:

$$\left.\begin{aligned} &\Delta \mathbf{u} -\lambda\frac{\partial \mathbf{u}}{\partial x_1} = \nabla\pi+F \\ &\nabla\cdot\mathbf{u} = 0 \end{aligned}\right\} \text{ in } \Omega$$
$$\mathbf{u} = \mathbf{v}_* - \mathbf{v}_\infty \quad \text{at } \partial\Omega \tag{2.25}$$
$$\lim_{|x|\to\infty} \mathbf{u}(x) = 0 ,$$

and so (2.24) along with the result of [G, Theorems 3.1, 3.2] and the assumptions on $\mathbf{v}_*$ allows us to deduce

$$\mathbf{u}\in D^{12/7}(\Omega), \ \partial\mathbf{u}/\partial x_1 \in L^{6/5}(\Omega) \tag{2.26}$$

and

$$p\in D^{1,6/5}(\Omega).$$

From Lemma 2.1, from [G, Theorem 1.1] and from this latter condition it also follows

$$p\in L^2(\Omega), \tag{2.27}$$

where, for simplicity, we put $p_0=0$. Therefore, by Lemma 2.1, by assumption and by (2.22), (2.26), (2.27) we find that (2.21) is verified for all $r\geq3$, $s\geq6/5$, $t\geq12/7$ and $\sigma\geq2$ and so, to prove the lemma completely, it remains to show it for $r\in(2,3)$, $s\in(1,6/5)$, $t\in(4/3,12/7)$ and $\sigma\in(3/2,2)$. To this end, we observe that being

$$\|\mathbf{u}\cdot\nabla\mathbf{u}\|_{12/11} \leq \|\mathbf{u}\|_3 |\mathbf{u}|_{12/7},$$

again from (2.25), from [G, Theorems 3.1, 3.2] and from the hypothesis on $\mathbf{v}_*$ we deduce, in particular,

$$\mathbf{u}\in D^{1,t}(\Omega), \qquad \text{for all } t\geq3/2. \tag{2.28}$$

Having this established, we notice

$$\|\mathbf{u}\cdot\nabla\mathbf{u}\|_q \leq \|\mathbf{u}\|_3 |\mathbf{u}|_{1,3q/(3-q)}$$

and, consequently, from the assumption on $\mathbf{u}$ and from (2.28) we recover

$$F\in L^q(\Omega), \qquad \mathbf{v}_*\in W^{2-1/q,q}(\partial\Omega),$$

for all $q\in(1,12/7)$. So, from [G, Theorems 3.1, 3.2] it follows

$$\mathbf{u}\in L^{4q/(4-2q)}(\Omega)D^{1,4q/(4-q)}(\Omega)$$
$$\frac{\partial \mathbf{u}}{\partial x_1} \in L^q(\Omega)$$
$$p\in L^{3q/(3-q)}(\Omega),$$

for all the above specified values of q. The proof of the lemma is, therefore, completed.

### 3. Summability Properties of *D*-solutions.

In this section we shall derive the summability properties (at large distances) of a *D*-solution and shall prove that, under suitable assumptions on the body force, they coincide with those of the fundamental Oseen tensor E, *cf.* (1.5)′, or, what amounts to the same thing, with that of corresponding solution of the linearized Oseen problem, *cf.* Theorem 3.1. This will be achieved by combining the results of Lemma 2.2 with those of the following lemma.

**Lemma 3.1.** *Let $\Omega$ be an exterior domain of class $C^2$ and assume*

$$f\in L^{6/5}(\Omega)\cap L^{3/2}$$

$$v_*\in W^{4/3,3/2}(\partial\Omega).$$

*Then, every D-solution to the Navier-Stokes problem in $\Omega$ satisfies the condition*

$$v-v_\infty\in L^3(\Omega).$$

*Proof.* By the result of [G, Remark 5.6] we have

$$u\equiv v-v_\infty\in L^4(\Omega),\ \partial v/\partial x_1\in L^{4/3}(\Omega). \tag{3.1}$$

Consider the sequence of problems, $k\in\mathbb{N}$,

$$\left.\begin{aligned} &\Delta w-\lambda\frac{\partial w}{\partial x_1}-u_k\cdot\nabla w = f+\nabla\pi \\ &\nabla\cdot w = 0 \end{aligned}\right\}\ \text{in}\ \Omega$$

$$w=v_*-v_\infty\equiv w_*\quad \text{at}\ \partial\Omega \tag{3.2}$$

$$\lim_{|x|\to\infty} w(x) = 0,$$

where $\{u_k\}$ is the sequence of functions associated to u by Lemma 1.1 and and corresponding to some $\varepsilon$ which will be specified later on in the formula below (3.6). As we have already observed [G, footnote 8], by the Sobolev inequality it is

$$f\in L^{6/5}(\Omega)\ \text{implies}\ f\in D_0^{-1,2}(\Omega);$$

also, by the trace embeddings, *cf.* Nečas (1967, Chapitre.2 Théorème 5.5 and

5.8), we find

$$w_* \in W^{1/2,2}(\partial\Omega).$$

Therefore, using, for instance, the Galerkin method, it is not hard to show, for each $k \in \mathbb{N}$, the existence of a *D*-solution $w$ to (3.2)[3]. However, setting

$$F = f + u_k \cdot \nabla w,$$

from the inequality

$$\|u_k \cdot \nabla w\|_{6/5} \le \|u_k\|_3 |w|_{1,2} ,$$

and, from $(1.10)_1$ of Lemma 1.1, we recover

$$F \in L^{6/5}(\Omega).$$

This condition, along with the hypothesis on $w_*$ and in view of the results of [G, Theorem 3.1], ensures the existence of a solution $w_1$, $\pi_1$ to the Oseen problem:

$$\left.\begin{aligned} \Delta w_1 - \lambda \frac{\partial w}{\partial x_1} &= F + \nabla \pi_1 \\ \nabla \cdot w_1 &= 0 \end{aligned}\right\} \text{ in } \Omega$$
$$w_1 = w_* \quad \text{at } \partial\Omega, \tag{3.3}$$
$$\lim_{|x| \to \infty} w_1(x) = 0,$$

such that

$$w_1 \in L^3(\Omega) \cap D^{1,12/7}(\Omega) \cap D^{2,6/7}(\Omega),$$
$$\frac{\partial w_1}{\partial x_1} \in L^{6/5}(\Omega),$$
$$\pi_1 \in D^{1,6/5}(\Omega),$$

and verifying the estimate

---

[3] By this we mean, as usual, that $w$ belongs to $D^{1,2}(\Omega)$, it is weakly divergence free, assumes the value $w_*$ at $\partial\Omega$ (in the trace sense) and

$$\lim_{|x| \to \infty} \int_S |w(x)| = 0;$$

moreover, $w$ satisfies the identity

$$(\nabla w, \nabla \varphi) - (w, \frac{\partial \varphi}{\partial x_1}) - (u_k . \nabla \varphi, w) = [f, \varphi]$$

for all $\varphi \in \mathcal{D}(\Omega)$.

$$a_1\|w_1\|_3+\lambda\left\|\frac{\partial w_1}{\partial x_1}\right\|_{6/5}+a_2|w_1|_{1,12/7}+|w_1|_{2,6/5}+|\pi_1|_{1,6/5}$$
$$\le c(\|F\|_{6/5}+\|w_*\|_{7/6,6/5(\partial\Omega)}) \tag{3.4}$$

where $a_1$, $a_2$ are constants defined in [G, Theorem 3.1]. From interpolation it follows

$$w_1 \in D^{1,2}(\Omega)$$

and, consequently, $w_1$ and $w$ are both *D*-solutions to the Oseen problem (3.3) assuming the same data. Thus, from [G, Theorem 3.2] we infer

$$w=w_1, \quad \nabla\pi = \nabla\pi_1, \quad \text{a.e. in } \Omega. \tag{3.5}$$

We next observe that, by the property (1.11) of the functions $u_k$, it is

$$\|u_k\cdot\nabla w\|_{6/5} \le \|u_k^{(1)}\|_4|w|_{1,12/7}+\|v-v_\infty\|_{\infty,\Omega_\rho}|w|_{1,6/5,\Omega_\rho}$$
$$<\varepsilon|w|_{1,12/7}+\|v-v_\infty\|_{\infty,\Omega_\rho}|w|_{1,6/5,\Omega_\rho} \tag{3.6}$$

and so, choosing

$$\varepsilon = a_2/(2c\lambda)$$

from (3.5), (3.6) it follows

$$\|w\|_3+\left\|\frac{\partial w}{\partial x_1}\right\|_{6/5}+|w|_{1,12/7}+|w|_{2,6/5}+|\pi|_{1,6/5}+\|\pi\|_2$$
$$\le c_1(\|f\|_{6/5}+\|w_*\|_{7/6,6/5,(\partial\Omega)}+|w|_{1,6/5,\Omega_\rho}) \tag{3.7}$$

Notice that the constant $c_1$ depends on $\Omega$, $\lambda$, $u$ but it is *independent* of k. Likewise, the radius $\rho$ depends on $\varepsilon$ and $u$ but it is otherwise independent of k. The next step is to show that there exists a positive constant $c_2=c_2(\Omega,\lambda,u,\rho)$ *independent of* k such that

$$|w|_{1,6/5,\Omega_\rho} \le c_2(\|f\|_{6/5}+\|w_*\|_{7/6,6/5,(\partial\Omega)}) \tag{3.8}$$

Contraddicting (3.8) means that there exist sequences

$$\{u_m\}, \quad \{f_m\}, \quad \{w_{*m}\}$$

such that, denoting by $\{w_m, \pi_m\}$ the sequence of corresponding solutions, it holds for all $m\in\mathbb{N}$

$$|w_m|_{1,6/5,\Omega_\rho} = 1,$$
$$\|f_m\|_{6/5}+\|w_{*m}\|_{7/6,6/5,(\partial\Omega)} \le \frac{1}{m} \tag{3.9}$$

From (3.7), (3.9) we immediately obtain the existence of $M>0$ independent of m such that

$$\|w_m\|_3+\left\|\frac{\partial w_m}{\partial x_1}\right\|_{6/5}+|w_m|_{1,12/7}+|w_m|_{2,6/5}+|\pi_m|_{1,6/5}\le M.$$

From this inequality and from weak compactness of the spaces $L^s(\Omega)$ we deduce the existence of a subsequence- denoted again by $\{w_m,\pi_m\}$- and of two fields $\mathbf{W}$, $\Pi$ such that

$$\begin{aligned}&\mathbf{W}\in L^3(\Omega)\cap D^{1,12/7}(\Omega)\cap D^{2,6/5}(\Omega),\\ &\frac{\partial \mathbf{W}}{\partial x_1}\in L^{6/5}(\Omega),\\ &\Pi\in D^{1,6/5}(\Omega),\end{aligned} \tag{3.10}$$

and

$$\begin{aligned}&w_m\to \mathbf{W}\quad \text{weakly in } W^{2,6/5}(\Omega_R)\\ &w_m\to \mathbf{W}\quad \text{strongly in } W^{1,6/5}(\Omega_R)\end{aligned} \tag{3.11}$$

where, in deriving $(3.11)_2$, we have employed the well known property that the embedding $W^{2,s}(A)\subset W^{1,s}(A)$ is compact for $A$ bounded and $s\ge 1$. By $(3.11)_2$ and $(3.9)_1$, we also have

$$|\mathbf{W}|_{1,6/5,\Omega_\rho}=1 \tag{3.12}$$

On the other hand, from (1.10) we derive

$$\|u_m\|_4\le B$$

for a constant $B$ independent of m. Thus, again by the weak compactness of $L^s$ spaces, we obtain (along a subsequence)

$$u_m\to U\quad \text{weakly in } L^4(\Omega) \tag{3.13}$$

for some $U\in L^4(\Omega)$. Since $\nabla\cdot u_m=0$ for all $m\in\mathbb{N}$, we have also $\nabla\cdot U=0$ (in the weak sense). Now, from (3.2) we derive, for each $m\in\mathbb{N}$,

$$(\nabla w_m,\nabla\varphi)-(w_m,\frac{\partial\varphi}{\partial x_1})-(u_m\cdot\nabla\varphi,w_m)=-[f_m,\varphi] \tag{3.14}$$

with $\varphi$ arbitrary from $\mathcal{D}(\Omega)$. Taking into account that

$$\nabla\varphi\cdot \mathbf{W}\in L^{4/3}(\Omega),$$

from (3.11), (3.13) (3.14) and $(3.9)_2$ it is not hard to show that the limiting field $\mathbf{W}$ obeys the identity

$$(\nabla W,\nabla\varphi)+(\frac{\partial W}{\partial x_1},\varphi)-(U\cdot\nabla\varphi,W) = 0, \tag{3.15}$$

for all $\varphi\in\mathcal{D}(\Omega)$. Moreover, $W$ is divergence free and, because of $(3.9)_2$, it has also zero trace at the boundary. Since, by $(3.10)_1$ and by [G, Theorem 1.1] it is

$$W\in D^{1,2}(\Omega),$$

from the characterization given in [G, (1.1)] we obtain

$$W\in\mathcal{D}_0^{1,2}(\Omega). \tag{3.16}$$

This property furnishes, in particular, by Sobolev inequality,

$$W\in L^6(\Omega)$$

which, together with $(3.10)_1$, yields by interpolation

$$W\in L^4(\Omega). \tag{3.17}$$

Moreover, again by (3.10) and by simple interpolation, we find

$$\frac{\partial W}{\partial x_1}\in L^{4/3}(\Omega). \tag{3.18}$$

With (3.16)-(3.18) in hands and recalling that $U\in L^4(\Omega)$ and $\nabla\cdot U=0$, we may procede exactly as in [G, Lemma 5.2] to show $W=0$. Let us briefly sketch the proof. By [G, Theorem 2.2] we may find a sequence $\{\varphi_k\}\subset\mathcal{D}(\Omega)$ such that

$$\varphi_k\to W, \quad \text{strongly in } \mathcal{D}_0^{1,2}(\Omega)\cap L^4(\Omega).$$

From (3.15) it then follows

$$|W|_{1,2}+(\frac{\partial W}{\partial x_1},W)-(U\cdot\nabla W,W) = 0. \tag{3.19}$$

Since $U$ is divergence free in the weak sense, from [G, Lemma 5.1] we have

$$(U\cdot\nabla W,W) = 0, \tag{3.20}$$

while, proceeding exactly as in [G, Lemma 5.2, the reasoning following eq. (5.12)] we show

$$(\frac{\partial W}{\partial x_1},W) = 0.$$

This condition, along with (3.19) and (3.20), entails

$$|W|_{1,2} = 0$$

which contradicts (3.12). We conclude that (3.8) holds and then, combining (3.8) and (3.7), that $w,\pi$ satisfy the following inequality

$$\|w\|_3+\left\|\frac{\partial w}{\partial x_1}\right\|_{6/5}+|w|_{1,12/7}+|w|_{2,6/5}+|\pi|_{1,6/5}+\|\pi\|_2$$
$$\le c_3(\|f\|_6+\|w_*\|_{7/6,6/5(\partial\Omega)}). \tag{3.21}$$

for a constant $c_3$ independent of W, $\pi$, f, $w_*$ and k. We, now, wish to let $k\to\infty$ into (3.2). For convenience, we label with the index k the corresponding solution $w_k$, $\pi_k$. We know that, for all $k\in\mathbb{N}$, $w_k,\pi_k$ obey (3.21) and so, reasoning as before, we obtain a pair $w,\pi$ enjoying properties of the type (3.10), (3.11) and (3.16)-(3.18). Clearly, $w_*$ is the trace of w at the boundary. Moreover, $w_k$ satisfies, for each $k\in\mathbb{N}$, the identity (3.14) written with k in place of m and f in place og $f_m$. And so, passing to the limit $k\to\infty$ into such relation and recalling $(1.10)_2$, we conclude that $w,\pi$ satisfy, a.e. in $\Omega$, the following system

$$\begin{aligned}&\Delta w-\lambda\frac{\partial w}{\partial x_1}-v\cdot\nabla w=f+\nabla\pi,\\&\nabla\cdot w=0,\end{aligned} \tag{3.22}$$

and also

$$\begin{aligned}&w=w_* \quad \text{at } \partial\Omega\\&\lim_{|x|\to\infty}\int_S |w(x)| = 0.\end{aligned} \tag{3.23}$$

Obviously, v is a solution to (3.22), (3.23) and it is not difficult to show that v=w, a.e. in $\Omega$. Actually, setting z=v−w, we deduce that z is a solution to (3.22), (3.23) with $w_*$=f=0. However, by (3.1) and by the properties (3.16)-(3.18) of w we have

$$\begin{aligned}&z\in\mathcal{D}_0^{1,2}(\Omega)\cap L^4(\Omega),\\&\frac{\partial z}{\partial x_1}\in L^{4/3}(\Omega).\end{aligned} \tag{3.24}$$

Since z solves

$$(\nabla z,\nabla\varphi)+(\frac{\partial z}{\partial x_1},\varphi)-(v\cdot\nabla\varphi,z)=0,\quad \text{for all } \varphi\in\mathcal{D}(\Omega),$$

using (3.24) and reasoning as we did previously to show W≡0 we arrive at z≡0 a.e. in $\Omega$ which completes the proof of the lemma.

On the strenght of this result and of Lemma 2.2 we thus have

**Theorem 3.1.** *Let $\Omega$ be a $C^2$-smooth, exterior domain and assume*

$$f\in L^q(\Omega),$$
$$v_*\in W^{2-1/q_0,1/q_0}(\partial\Omega),$$

*for some* $q_0>3$, *and all* $q\in(1,q_0]$. *Then every D-solution* $v$ *to the Navier-Stokes problem corresponding to* $f$, $v_*$ *and* $v_\infty$ *satisfies*

$$v-v_\infty\in L^r(\Omega),$$
$$\frac{\partial v}{\partial x_1}\in L^s(\Omega),$$
$$v\in D^{1,t}(\Omega),$$
$$p\in L^\sigma(\Omega),$$

*for all* $r\in(2,\infty]$, $s\in(1,\infty]$, $t\in(4/3,\infty]$, *where* $p$ *(up to a constant) is the pressure field associated to* $v$.

**4. Asymptotic Structure of *D*-solutions.**

In this final section we determine the asymptotic structure of a *D*-solutions. Specifically, we shall prove that *D*-solutions posses the *same* structure as that corresponding to the linearized Oseen system, *cf.* (0.1). This, in particular, implies that *D*-solutions exhibit, at large dinstances, a paraboloidal wake region beyond $\Omega^c$.

Our first objective is to recover an appropriate uniform estimate for $u(x)=v(x)-v_\infty$, for large values of $|x|$. For simplicity, we shall assume that the body force $f$ is of bounded support. We recall the following notation:

$$B_a(x) = \{x\in R^3: |x-y|<a\},\quad B_a\equiv B_a(0)\quad (a>0).$$

Let $\Omega$ denote, temporarily, the exterior of $B_\rho$, where $\rho$ is taken as large as to satisfy $B_\rho\supseteq supp(f)$. By [G, Theorem 4.2] we then have that $v,p\in C^\infty(\Omega)$ and that, by Theorem 3.1, $v,p$ enjoy the summability properties (2.21). Consider, next, the representation formula (1.6), $(1.8)_1$ (1.9) in $\Omega$. From Theorem 1.2, we have

$$u(x) = N[u(x)]+O(1/|x|)\quad \text{as } |x|\to\infty. \tag{4.1}$$

where

$$N_j=(N[u(x)])_j = \int_{\Omega_R} E_{ij}(x-y)u(y)\cdot\nabla u_i(y)dy.$$

We wish to give a uniform estimate for N. To this end, setting $|x|=2R$, we may write

$$N_j=\int_{\Omega_R} E_{ij}(x-y)u(y)\cdot\nabla u_i(y)dy+\int_{\Omega^R} E_{ij}(x-y)u(y)\cdot\nabla u_i(y)dy \equiv \mathcal{N}_1+\mathcal{N}_2. \tag{4.2}$$

Recalling that

$$|E_{ij}(x-y)| \le c|x-y|^{-1},$$

*cf.* $(1.5)_1$, it follows

$$|\mathcal{N}_1| \le \frac{c}{R}\|u\|_{3,\Omega_R}|u|_{1,3/2,\Omega_R} \le \frac{2c}{|x|}\|u\|_3|u|_{1,3/2} \tag{4.3}$$

Inequality (4.3) and Theorem 3.1 yield

$$|\mathcal{N}_1| \le c_1|x|^{-1}. \tag{4.4}$$

Furthermore, from (1.2) and $(1.5)_1$ it is also

$$|\mathcal{N}_2| \le \left(\int_{\Omega^R}\frac{u^2}{|x-y|^2}\right)^{1/2}\left(\int_{\Omega^R}\nabla u:\nabla u\right)^{1/2} \le c_2\int_{\Omega^R}\nabla u:\nabla u. \tag{4.5}$$

with $c_2$ independent of u and R. Setting

$$G(R) \equiv \int_{\Omega^R}\nabla u:\nabla u \tag{4.6}$$

from (4.1), (4.3) and (4.4) it follows that estimating the nonlinear term $N[u(x)]$ is reduced to estimating the functions $G(R)$, $R=|x|/2$. This latter question will be analyzed in the next two lemmas. We begin with a simple but very useful result.

**Lemma 4.1.** *Suppose that, for all* $t\ge a>0$,

(*i*) $y(t)\in C^1(t,\infty)$.

(*ii*) $y(t) \ge 0$,

*iii*) $y'(t)\le 0$,

*and that, for some* $\beta\in[0,1]$,

$$\int_a^\infty y(s)s^{1-\beta}ds<\infty.$$

*Then, for all* $t\ge a$ *it is*

$$y(t)\,t^{1-\beta} \le (1-\beta)\int_a^\infty y(s)ds + y(a)a^{1-\beta}.$$

*Proof*. The assertion is an immediate consequence of the identity

$$y(t)\,t^{1-\beta} = \int_a^\infty \frac{d}{ds}[y(s)s^{1-\beta}]ds + y(a)a^{1-\beta} \tag{4.7}$$

*and of the assumptions made on y.*

Lemma 4.1 allows us to prove the following estimate for $G(R)$.

**Lemma 4.2** - *Let* v *be a D-solution to the Navier-Stokes equations in* $\Omega=\mathbb{R}^3-B_\rho$ *corresponding to* f=0. *Then, setting* $u=v-v_\infty$, *for all* $R>\rho$ *it holds*

$$G(R) \le cR^{-1+\varepsilon}$$

*where* G *is defined in* (4.6), ε *is an arbitrary positive number and* c *is independent of* R.

*Proof.* As already observed, by the result of [G, Theorem 4.2] we may assume v and the correponding pressure p to be in $C^\infty(\Omega)$. Multiplying (0.1) by u and integrating over $\Omega_{R,R^*}$ $(R>R^*)$ it follows

$$\int_{\Omega_{R,R^*}} \nabla u:\nabla u = \int_{\partial B_R \cup \partial B_{R^*}} \left\{u\cdot\frac{\partial u}{\partial n} - \lambda|u|^2 v\cdot n - p(u\cdot n)\right\} \tag{4.8}$$

From Theorem 3.1 we derive, in particular,

$$\Psi \equiv |u|^3+|\nabla u|^{3/2}+|u|^{5/2}+|p|^5 \in L^1(\Omega). \tag{4.9}$$

Therefore, there exists a sequence $\{R_k\}$ tending to infinity as k tends to infinity, such that

$$\int_{\partial B_{R_k}} \Psi(R_k,\omega)d\omega = o(R_k^{-1}) \tag{4.10}$$

Using Young's inequality several times and Hölder inequality, we deduce (with r denoting either R or $R^*$)

$$\begin{aligned} F(r) &\equiv \int_{\partial B_r} \left[u\cdot\frac{\partial u}{\partial n} - \mathcal{R}e|u|^2 v\cdot n - p(u\cdot n)\right] \\ &\le c\left\{\int_{\partial B_r}\left[u^3+|\nabla u|^{3/2}+u^{5/2}+p^5\right] + r^{2/q}\left(\int_{\partial B_r} u^{2q}\right)^{1/q}\right\} \end{aligned} \tag{4.11}$$

where q is an arbitrary number greater than one. Taking in (4.11) $r=R_k$, q=5/2 and using (4.9), (4.10) it follows

$$\lim_{k\to\infty} F(R_k) = 0,$$

and (4.7) furnishes

$$G(R) = F(R) \tag{4.12}$$

For any $\varepsilon\in(0,1)$, by Young's inequality, it is

$$h(R) = R^{-\varepsilon}R^{2/q'}\left(\int_{\partial B_r} u^{2q}\right)^{1/q}\Big\} \le c\left[R^{-\varepsilon q'+2}+\int_{\partial B_r} u^{2q}\right]$$

and so, taking $q>3/(3-\varepsilon)$, and recalling that, by Theorem 3.1, it is $u\in L^{s}(\Omega)$ for all $s>2$, we have

$$h\in L^{1}(\rho,\infty). \tag{4.13}$$

Collecting (4.9), (4.11) and 4.13) yields

$$R^{-\varepsilon}F(R)\in L^{1}(\rho,\infty) \tag{4.14}$$

and, since

$$G'(R) = -\int_{\partial B_r} \nabla u:\nabla u < 0, \tag{4.15}$$

From (4.12), (4.14) and (4.15), with the help of Lemma 4.1, we prove the assertion of the lemma.

Collecting (4.2), (4.4), (4.5) together with the results of lemma 4.2 we then have the following uniform estimate for the non-linear term:

$$|N[u(x)]| \le c|x|^{-1+\varepsilon}, \quad \text{any } \varepsilon\in(0,1]. \tag{4.16}$$

From (4.1) and (4.16) we thus obtain

**Lemma 4.3** - *Let* $v$ *be a D-solution to the Navier-Stokes equation in a three dimensional exterior domain* $\Omega$*, corresponding to* $f$ *of bounded support. Then, for all sufficiently large* $|x|$*, the following uniform estimate holds:*

$$v(x) = v_\infty + O(1/|x|^{1-\varepsilon}), \quad \textit{any } \varepsilon\in(0,1].$$

With this result in hands, we are now able to show the following theorem which furnishes the asymptotic structure of any generalized solution corresponding to a body force of bounded support.

**Theorem 4.1** - *Let* $v$ *be a D-solution to the Navier-Stokes problem in a three-dimensional, exterior domain of class* $C^2$*, corresponding to* $f$*,* $v_*$ *where* $f$ *is of bounded support in* $\Omega$ *and, moreover,*

$$f\in L^{q}(\Omega),$$
$$v_*\in W^{2-1/q_o,1/q_o}(\partial\Omega),$$

*for some* $q_o>3$ *and all* $q\in(1,q_o]$*. Then, for all sufficiently large* $|x|$*,* $v$ *admits the following representation*

$$v(x) = v_\infty + m\cdot E(x) + \mathcal{T}(x) \tag{4.17}$$

*where E is the Oseen fundamental tensor,*

$$m = -\int_{\partial\Omega}\Big[T(v,p)\cdot n - \lambda(v-v_\infty)\cdot n\Big] + \int_\Omega f \qquad (4.18)$$

*and* $\mathcal{T}(x)$ *satisfies the estimate*

$$\mathcal{T}(x) = O(|x|^{-3/2+\delta}), \quad \text{any } \delta>0. \qquad (4.19)$$

*Proof.* In view of Theorem 1.2, to show the result it suffices to show the uniform estimate

$$\Big|\int_\Omega D_\ell E_{ij}(x-y)u_\ell(y)u_i(y)dy\Big| = O(|x|^{-3/2+\delta}). \qquad (4.20)$$

Setting $|x|=R$, sufficiently large, we divide the region $\Omega$ into three parts:

$$\Omega_{R/2},\ \Omega_{R/2,2R},\ \Omega^{2R}$$

and denote the corresponding iontegrals over these regions by $I_1$, $I_2$, $I_3$. Using $(1.5)_1$ and Hölder inequality, we find

$$|I_1| \le \frac{c}{R^{3/2}}\int_{\Omega_{R/2}} u^2 \le c_1 R^{[(3/q')-(3/2)]}\|u\|^2_{2q}$$

and so, choosing $q=3/(3-\delta)$, from Theorem 3.1 it follows

$$|I_1| \le c_2|x|^{-3/2+\delta}. \qquad (4.21)$$

Furthermore, recalling Lemma 4.3, we have

$$|I_2| \le c_3R^{-2+2\varepsilon}|E|_{1,2,\Omega_{R/2,2R}} \le c_3R^{-2+2\varepsilon}|E|_{1,2,B_{3R}(x)}$$

and so, in view of $(1.5)_2$, choosing $\varepsilon=\delta/2$, we recover

$$|I_2| \le c_4|x|^{-3/2+\delta} \qquad (4.22)$$

Finally, again from lemma 4.3 and $(1.5)_2$, it follows

$$|I_3| \le c_5\int_{2R}^{\infty} r^{-2+2\varepsilon}r^{-1/2}dr \le c_6R^{-3/2+2\varepsilon}$$

which by the chioce $\varepsilon=\delta/2$, in turn implies

$$|I_3| \le c_7|x|^{-3/2+\delta}. \qquad (4.23)$$

The validity of (4.20) is a consequence of (4.21)-(4.23) and the theorem is completely proved.

Finally, we wish to notice that estimates analogous to (4.20) can be shown for the first derivatives of v and for the pressure field p. We shall not consider this here and refer the interested reader to the book of Galdi (forthcoming).

## References

BABENKO, K.I., 1973, On Stationary Solutions of the Problem of Flow Past a Body of a Viscous Incompressible Fluid, *Mat. Sbornik,* 91 (133), p.1. English Translation in *Math.* USSR *Sbornik* 20 (1973), p.1.

BOGOVSKIĬ, M.E., 1980, Solution of Some Vector Analysis Problems Connected with Operators Div and Grad, *Trudy Seminar S.L.Sobolev,* #1, 80, *Akademia Nauk SSSR, Sibirskoe Otdelnie Matematiki, Nowosibirsk,* p.5 (in Russian).

FINN, R., 1959, Estimates at Infinity for Stationary Solutions of the Navier-Stokes Equations, *Bull. Math. Soc. Sci. Math. Phys. R.P. Roumaine,* 3 (51), p. 387

FINN, R., 1965, On the Exterior Stationary Problem for the Navier-Stokes Equations, and Associated Perturbation Problems, *Arch. Ratl Mech. Anal.* 19, p.363.

FUJITA, H., 1961, On the Existence and Regularity of the Steady-State Solutions of the Navier-Stokes Equations, *J.Fac.Sci.Univ.Tokyo,* Sect. I, 9, p.59.

GALDI, G.P., 1992, On the Oseen Boundary-Value Problem in Exterior Domains, *Proc. of the Oberwolfach Meeting "The Navier-Stokes Equations: Theory and Numerical Methods", J.G.Heywood, K.Masuda, R.Rautmann and V.A.Solonnikov Eds., Springer-Verlag Lecture Notes in Mathematics,* in the Press.

GALDI, G.P. (*forthcoming*) An Introduction to the Mathematical Theory of the Navier-Stokes Equations, *Springer Tracts in Natural Phylosophy.*

GALDI, G.P. and Maremonti, P., 1986, Monotonic Decreasing and Asymptotic Behavior of the Kinetic Energy for Weak Solutions of the Navier-Stokes Equations in Exterior Domains, *Arch. Ratl Mech. Anal.*, 94, p.253

NEČAS, J., 1967, Les Méthodes Directes en Théorie des Équations Elliptiques, *Masson et C*[ie].

OSEEN, C.W., 1927, Neuere Methoden Und Ergebnisse in Der Hydrodynamik, *Leipzig, Akademische Verlagsgesellschaft M.B.H.*.

# On the Solvability of an Evolution Free Boundary Problem for the Navier-Stokes Equations in Holdër Spaces of Functions.

I. S. Mogilevskii
Math. Fac. Univ. Of Kalinin
170 013 Kalinin

V. A. Solonnikov
Math. Inst. of V.A. Steklov
191 01 Petersburg

*To Professor G. Prodi on his 65-th birthday.*

We consider the following free boundary problem governing the motion of an isolated liquid mass: find a bounded domain $\Omega_t \subset \mathbb{R}^3$, $t > 0$, the vector field of velocities $\underset{\sim}{v}(x,t) = (v_1,v_2,v_3)$ and a scalar pressure $p(x,t)$ satisfying the Navier-Stokes equations

$$\underset{\sim}{v}_t + (\underset{\sim}{v}\cdot\nabla)\underset{\sim}{v} - \nu\nabla^2\underset{\sim}{v} + \nabla p = \kappa\nabla V + \underset{\sim}{f}(x,t), \qquad (1.1)$$

$$\nabla\cdot\underset{\sim}{v} = 0 \qquad (x \in \Omega_t,\ t > 0)$$

and initial and boundary conditions

$$\underset{\sim}{v}(x,0) = \underset{\sim}{v}_0(x) \qquad (x \in \Omega_0 \equiv \Omega), \qquad (1.2)$$

$$\mathbb{T}\underset{\sim}{n} - \sigma H\underset{\sim}{n} = 0 \qquad (x \in \Gamma_t \equiv \partial\Omega_t,\ t > 0) \qquad (1.3)$$

Here

$$\nabla = \left(\frac{\partial}{\partial x_1}, \frac{\partial}{\partial x_2}, \frac{\partial}{\partial x_3}\right), \quad \nabla\cdot\underset{\sim}{v} = \operatorname{div} \underset{\sim}{v}, \quad \nabla p = \operatorname{grad} p,$$

$\nu$ and $\sigma$ are positive constant coefficients of the viscosity and of the surface tension (we have assumed that the density of the liquid equals 1);

$$V(x,t) = \int_{\Omega_t} \frac{dy}{|x - y|}$$

is a gravitational potential, so that $\kappa\nabla V$ is a Newtonian attraction force acting on a liquid particle at the point x, $\underset{\sim}{f}(x,t)$ is a given vector field of external forces, $\mathbb{T} = -p\mathbb{I} + \nu\mathbb{S}(\underset{\sim}{v})$ is the stress tensor, $\mathbb{S}(\underset{\sim}{v})$ is the strain tensor with the elements

$$S_{ij}(v) = \frac{\partial v_i}{\partial x_j} + \frac{\partial v_j}{\partial x_i} ,$$

$\underset{\sim}{n}(x,t)$ is the exterior unit normal vector to the surface $\Gamma_t$ at the point x, and H(x,t) is the twice mean curvature of $\Gamma_t$ at the point x, which is supposed to be negative if Ω is convex in the neighbourhood of x. With this choice of the sign of H, there holds

$$H\underset{\sim}{n} = \Delta(t)\underset{\sim}{x}$$

where Δ(t) is the Laplace-Beltrami operator on $\Gamma_t$ and $\underset{\sim}{x}$ is the radius-vector corresponding to the point x ($x \in \Gamma_t$).

In addition to (1.3), we prescribe a kinematic boundary condition which says that $\Omega_t$ is the set of points $x = x(t) \subset R^3$ such that

$$\frac{d\underset{\sim}{x}(\tau)}{d\tau} = \underset{\sim}{v}(x(\tau),\tau) \qquad (0 \le \tau \le t) \tag{1.4}$$

$$\underset{\sim}{x}(0) = \underset{\sim}{\xi}$$

and ξ is an arbitrary point of Ω. $x = x(t)$ and $\xi = x(0)$ are usually referred to as the Eulerian and the Lagrangian coordinates of a liquid particle.

Let $\underset{\sim}{u}(\xi,t)$ and $q(\xi,t)$ be the velocity and the pressure expressed as function of Lagrangian coordinates. Then (1.4) is easily integrated:

$$\underset{\sim}{x}(t) = \underset{\sim}{\xi} + \int_0^t \underset{\sim}{u}(\xi,\tau)\, d\tau \equiv X_u(\xi,\tau), \tag{1.5}$$

and (1.1) - (1.3) takes a form of an initial boundary value problem in a given domain Ω, namely,

$$\begin{aligned} &\underset{\sim}{u}_t - \nu\nabla_u^2\underset{\sim}{u} + \nabla_u q = \kappa\nabla_u V(X_u,t) + \underset{\sim}{f}(X_u,t), \\ &\nabla_u\cdot\underset{\sim}{u} = 0 \qquad\qquad (\xi \in \Omega,\ t > 0), \\ &\underset{\sim}{u}(\xi,0) = \underset{\sim}{v}_0(\xi) \qquad (\xi \in \Omega), \\ &\mathbb{T}_u\underset{\sim}{n} - \sigma\Delta(t)X_u(\xi,t) = 0 \quad (\xi \in \Gamma = \partial\Omega). \end{aligned} \tag{1.6}$$

Here

$$\nabla_u = \mathcal{A}\nabla = \left(\sum_{m=1}^{3} A_{im} \frac{\partial}{\partial\xi_m}\right)_{i=1,2,3} \quad \text{with } A_{im} = \frac{\partial\xi_m}{\partial x_i},$$

but in virtue of incompressibility condition $\nabla_u\cdot\underset{\sim}{u} = \nabla\cdot\underset{\sim}{v} = 0$, the Jacobian

$$\det\left(\frac{\partial X_u}{\partial \xi}\right)$$

equals 1 and $A_{lm}$ can be computed as a cofactor of the element

$$a_{lm} = \delta_{lm} + \int_0^t \frac{\partial u_l}{\partial \xi_m}\, d\tau$$

of the Jacobi matrix $\frac{\partial X_u}{\partial \xi}$. The vector $\underset{\sim}{n}(X_u,t)$ and exterior normal vector $\underset{\sim}{n}_0(\xi)$ to $\Gamma$ at the point $\xi$ are related to each other by the formula

$$\underset{\sim}{n} = \frac{\mathcal{A}\underset{\sim}{n}_0}{|\mathcal{A}\underset{\sim}{n}_0|};$$

finally ,

$$\mathbb{T}_u = -q\mathbb{I} + \nu\mathbb{S}_u(\underset{\sim}{u}),\ (S_u(\underset{\sim}{w}))_{ij} = \sum_{m=1}^{3}\left(A_{mj}\frac{\partial u_i}{\partial \xi_m} + A_{mi}\frac{\partial u_i}{\partial \xi_m}\right).$$

Problem (1.1) - (1.4) was studied in a series of papers of V.A. Solonnikov [S 7–12] where it was proved that it has a unique solution in a certain finite time interval (0,T') (with T' depending on the data) for arbitrary $\Omega_0 \equiv \Omega$ and arbitrary $\underset{\sim}{v}_0$ possessing some regularity properties and satisfying natural compatibility conditions. However, if $\underset{\sim}{f} = 0$ and initial data are close to the equilibrium state, i.e. $\underset{\sim}{v}_0$ is small and $\Omega$ is close to a ball, then the solution exists for all $t \ge 0$, and as $t \to \infty$ , the solution tends to a limiting regime corresponding to the rotation of the liquid as a rigid body with a constant angular speed about a certain axis also moving with a constant speed. The case $\sigma = 0$ is treated in [S 6,10].

Similar problem describing an unsteady motion of a viscous incompressible liquid in an infinite layer between a rigid bottom and a free surface was considered by Th. Beale [B 1,2] and G. Allain [A] who proved local solvability of the problem for arbitrary initial data and global solvability for small initial data.

Y. Teramoto [T 1,2] studied unsteady viscous flow down an inclined plane (both for $\sigma = 0$ and for $\sigma > 0$) and established an existence theorem for the corresponding problem on a finite time interval; in the case $\sigma > 0$ the interval may be arbitrary.

In all these papers, with the exception of [S6], problem (1.1) - (1.4) was investigated in Sobolev spaces of functions. However, it may be preferable to work in Hölder spaces, especially when $\sigma$ is not a constant, as in the problem of thermocapillary convection treated by M.V. Lagunova and V.A. Solonnikov [LS].

In the present paper, we prove a local existence theorem for the problem (1.6) in Hölder spaces. This implies the Hölder continuity of derivatives of global solution obtained in [S 8,11], provided the data satisfy appropriate regularity hypotheses. These results are announced [MS 2].

In addition to (1.6), we consider an auxiliary linear problem

$$
\begin{aligned}
&\boldsymbol{w}_t - \nu\nabla_u^2 \boldsymbol{w} + \nabla_u s = \boldsymbol{f}(\xi,t),\\
&\nabla_u \cdot \boldsymbol{w} = \rho(\xi,t) \qquad (\xi \in \Omega,\ t > 0),\\
&\boldsymbol{w}(\xi,0) = \boldsymbol{w}_0(\xi),\\
&\nu\Pi_0\Pi S_u(\boldsymbol{w})\boldsymbol{n} = \Pi_0\boldsymbol{d},\\
&\boldsymbol{n}_0\cdot T_u(\boldsymbol{w},s)\boldsymbol{n} - \sigma(\xi)\boldsymbol{n}_0\cdot\Delta(t)\int_0^t \boldsymbol{w}\,d\tau = b + \int_0^t B\,d\tau.
\end{aligned}
\tag{1.7}
$$

Here operator $\nabla_u = \mathcal{A}\nabla$, matrices $S_u(\boldsymbol{w})$, $T_u(\boldsymbol{w},s)$ and vector $\boldsymbol{n}$ are determined by a given vector field $\boldsymbol{u}$, as explained above, and $\Delta(t)$ is the Laplace-Beltrami operator on the surface $\Gamma_t = X_u\Gamma$. By $\Pi_0$ and $\Pi$ we mean projection operators onto tangent planes to $\Gamma$ and $\Gamma_t$ respectively, i.e.

$$\Pi_0\boldsymbol{g} = \boldsymbol{g} - \boldsymbol{n}_0(\boldsymbol{n}_0\cdot\boldsymbol{g}), \quad \Pi\boldsymbol{g} = \boldsymbol{g} - \boldsymbol{n}(\boldsymbol{n}\cdot\boldsymbol{g}),$$

where $\boldsymbol{a}\cdot\boldsymbol{b} = a_1b_1 + a_2b_2 + a_3b_3$. We notice that the equality $\nu\Pi_0\Pi S_u(\boldsymbol{w})\boldsymbol{n} = \Pi_0\boldsymbol{d}$ contains only two independent boundary conditions.

In the case $\boldsymbol{u} = 0$ problem (1.7) takes a form

$$
\begin{aligned}
&\boldsymbol{w}_t - \nu\nabla^2\boldsymbol{w} + \nabla s = \boldsymbol{f}, \qquad \nabla\cdot\boldsymbol{w} = \rho \qquad (\xi \in \Omega,\ t > 0),\\
&\boldsymbol{w}(\xi,0) = \boldsymbol{w}_0(\xi) \qquad (\xi\in\Omega),\\
&\nu\Pi_0 S(\boldsymbol{w})\boldsymbol{n}_0 = \Pi_0\boldsymbol{d},\\
&\boldsymbol{n}_0 . T(\boldsymbol{w},s)\boldsymbol{n}_0 - \sigma(\xi)\boldsymbol{n}_0\cdot\Delta\int_0^t \boldsymbol{w}\,d\tau = b + \int_0^t B\,d\tau \quad (\xi \in \Gamma)
\end{aligned}
\tag{1.8}
$$

where $\Delta = \Delta(0)$ is the Laplace-Beltrami operator on $\Gamma$.

To describe our results more precisely, we need a definition of Hölder spaces. Let G be a domain in $\mathbb{R}^n$ and let r be a positive non-integer, i.e. $r = [r] + \{r\}$, $\{r\} \in (0,1)$. By $C^r(G)$ we mean the spaces of [r] times continuously differentiable funtions whose [r]-th derivatives satisfy Hölder condition with exponent {r}, i.e. $C^r(G)$ is the space of functions with a finite norm

$$|u|_G^{(r)} = \sum_{|k| \le [r]} |D^k u|_G + \langle u\rangle_G^{(r)}$$

where

$$D^k u = \frac{\partial^{|k|} u}{\partial x_1^{k_1} \dots \partial x_n^{k_n}}, \qquad |v|_G = \sup_{x \in G} |v(x)|$$

and

$$\langle u\rangle_G^{(r)} = \sum_{|k|=[r]} \sup_{x,y \in G} |x-y|^{-\{r\}} |D^k u(x) - D^k u(y)|.$$

For funtions depending on $(x,t) \in G \times (0,T) \equiv \mathcal{G}_T$, we define anisotropic Hölder space $C^{r_1,r_2}(\mathcal{G}_T)$ as the set of funtions with a finite norm

$$|u|_{C^{r_1,r_2}(\mathcal{G}_T)} = \sup_{t \in (0,T)} \left(|u(.,t)|_G^{(r_1)} + \sup_{x \in G} |u(x,.)|_{(0,T)}^{(r_2)}\right).$$

We shall deal almost exclusively with the case $r_2 = r_1/2$, and we prefer to work with another norm containing also mixed derivatives

$$D_x^k D_t^a u, 2|k| + a < r_1 .$$

Thus, we introduce in the space $C^{r,r/2}(\mathcal{G}_T)$ the norm

$$|u|_{\mathcal{G}_T}^{(r,r/2)} = \sum_{2a+|k| \le r} |D_x^k D_t^a u|_{\mathcal{G}_T} + \sum_{2a+|k|=[r]} \langle D_x^k D_t^a u\rangle_{\mathcal{G}_T}^{(\{r\})}$$

$$+ \sum_{2a+|k|=[r]-1} \langle D_x^k D_t^a u\rangle_{t,\mathcal{G}_T}^{\left(\frac{1+\{r\}}{2}\right)}$$

where

$$\langle v\rangle^{(\lambda)}_{\mathcal{G}_T} = \langle v\rangle^{(\lambda)}_{x,\mathcal{G}_T} + \langle v\rangle^{(\lambda/2)}_{t,\mathcal{G}_T},$$

$$\langle v\rangle^{(\lambda)}_{t,\mathcal{G}_T} = \sup_{t\in(0,T)} \langle v(\cdot,t)\rangle^{(\lambda)}_G = \sup_t \sup_{x,y} \frac{|v(x,t) - v(y,t)|}{|x - y|^\lambda},$$

$$\langle v\rangle^{(\mu)}_{x,\mathcal{G}_T} = \sup_{x\in G} \langle v(x,\cdot)\rangle^{(\mu)}_{(0,T)} = \sup_x \sup_{\tau,t\in(0,T)} \frac{|v(x,t) - v(x,\tau)|}{|t - \tau|^\mu}$$

for $\lambda,\mu \in (0,1)$. It can be shown that this norm is equivalent to

$$|u|_{C^{r,r/2}(\mathcal{G}_T)}$$

for any fixed T but the ratio

$$|u|^{r,r/2}_{\mathcal{G}_t} \left(|u|_{C^{r,r/2}(\mathcal{G}_T)}\right)^{-1}$$

may be unbouded as $T \to 0$.

For the estimates of the pressure we shall need also a seminorm

$$|u|^{(1+\alpha,\gamma)}_{\mathcal{G}_T} = \sup_{0<\tau<t<T} \tau^{-\frac{1+\alpha-\gamma}{2}} |u(\cdot,t) - u(\cdot,t-\tau)|^{(\gamma)}_\Omega$$

and

$$\langle u\rangle^{(1+\alpha,\gamma)}_{\mathcal{G}_T} = \sup_{0<\tau<t<T} \tau^{-\frac{1+\alpha-\gamma}{2}} \langle u(\cdot,t) - u(\cdot,t-\tau)\rangle^{(\gamma)}_\Omega$$

with $\alpha,\gamma \in (0,1)$.

All the above definitions extend in a natural way to vector valued funtions. The spaces of vector valued funtions will be denoted by the same symbols as spaces of scalar functions (for instance, we write $\underset{\sim}{v} \in C^1(\Omega)$ rather than $\underset{\sim}{v} \in (C^1(\Omega))^3$ etc). Using local maps and partition of unity, we may define Hölder spaces on manifolds (for example, on $\Gamma$ and on $\Gamma \times (0,T)$).

Now we can formulate our results. Consider first linear problems (1.7), (1.8). Let $Q_T = \Omega \times (0,T)$, $S_T = \Gamma \times (0,T)$ with arbitrary positive fixed $T > 0$ and let $\alpha \in (0,1)$. We assume that the data of linear problems possess the following properties:

i) $\underset{\sim}{f} \in C^{\alpha,\alpha/2}(Q_T)$, $\rho \in C^{1+\alpha,\frac{1+\alpha}{2}}(Q_T)$, $\underset{\sim}{v}_0 \in C^{2+\alpha}(\Omega)$,

$\underset{\sim}{d} \in C^{1+\alpha,\frac{1+\alpha}{2}}(S_T)$, $b \in C^{1+\alpha,\frac{1+\alpha}{2}}(S_T)$, $B \in C^{\alpha,\alpha/2}(S_T)$.

ii) The following compatibility conditions hold

$$\nabla . \underset{\sim}{v}_0 = \rho(\xi,0), \quad \nu\Pi_0 S(\underset{\sim}{v}_0)\underset{\sim}{n}_0\Big|_\Gamma = \Pi_0 \underset{\sim}{d}(\xi,0),$$

iii) There exist vector fields $\underset{\sim}{h} \in C^{\alpha,\alpha/2}(Q_T)$ and $\underset{\sim}{h}_k$, $k = 1,2,3$ continuous and with a finite norm $|\underset{\sim}{h}_k|^{(1+\alpha,\gamma)}_{Q_T}$, $\gamma \in (0,1)$ such that

$$\frac{\partial\rho}{\partial t} - \nabla_u \cdot \underset{\sim}{f} = \nabla\cdot\underset{\sim}{h}, \quad \underset{\sim}{h} = \sum_{k=1}^{3} \frac{\partial \underset{\sim}{h}_k}{\partial \xi_k} \tag{1.9}$$

This formulas should be understood in a distributional sense, i.e. as integral identities

$$\int_\Omega \rho(\xi,t)\varphi(\xi)d\xi - \int_\Omega \rho(\xi,0)\varphi(\xi)d\xi + \int_0^t\int_\Omega \underset{\sim}{f}(\xi,\tau)\cdot\nabla\mathcal{A}^*\varphi d\xi d\tau$$

$$= - \int_0^t\int_\Omega \underset{\sim}{h}(\xi,\tau)\cdot\nabla\varphi(\xi)d\xi d\tau \,,$$

$$\int_\Omega \underset{\sim}{h}(\xi,\tau)\varphi(\xi)d\xi = - \sum_{k=1}^{3} \int \underset{\sim}{h}_k(\xi,\tau)\varphi_{\xi_k}(\xi)d\xi$$

with arbitrary $\varphi \in C_0^\infty(\Omega)$.

In the problem (1.8) $\underset{\sim}{u} = 0$ and (1.9) take the form

$$\frac{\partial\rho}{\partial t} - \nabla\cdot\underset{\sim}{f} = \nabla\cdot\underset{\sim}{h} \,,\ \underset{\sim}{h} = \sum_{k=1}^{3} \frac{\partial \underset{\sim}{h}_k}{\partial \xi_k} \,. \tag{1.9'}$$

In §§ 3 and 4 we prove the following theorem.

**Theorem 1.1** *Suppose that* $\Gamma \in C^{2+\alpha}$, $\sigma \in C^\alpha(\Gamma)$, $\sigma \ge \sigma_0 > 0$. *Then for arbitrary data satisfying hypotheses i)–iii) problem (1.8) has a unique solution with the following differentiability properties :*

$$\underset{\sim}{w} \in C^{2+\alpha,1+\alpha/2}(Q_T),\ \nabla s \in C^{\alpha,\alpha/2}(Q_T),\ s \in C^{2+\alpha,1+\alpha/2}(S_T),\ |s|_{Q_T}^{(1+\alpha,\gamma)} < \infty,$$

*and the solution satisfies the inequality*

$$|\underset{\sim}{w}|_{Q_t}^{(2+\alpha,1+\alpha/2)} + |\nabla s|_{Q_t}^{\alpha,\alpha/2} + |s|_{S_t}^{\left(1+\alpha,\frac{1+\alpha}{2}\right)} + |s|_{Q_t}^{(1+\alpha,\gamma)}$$

$$\le C(t)(|\underset{\sim}{f}|_{Q_t}^{\alpha,\alpha/2} + |\underset{\sim}{w}_0|_{\Omega}^{2+\alpha} + |\nabla\rho|_{Q_t}^{(\alpha,\alpha/2)} + |\rho|_{Q_t}^{(1+\alpha,\gamma)}$$

$$+ |\rho|_{S_t}^{\left(1+\alpha,\frac{1+\alpha}{2}\right)} + |\underset{\sim}{h}|_{Q_t}^{(\alpha,\alpha/2)} + \sum_{k=1}^{3} |\underset{\sim}{h}_k|_{Q_t}^{(1+\alpha,\gamma)} + |\underset{\sim}{d}|_{S_t}^{\left(1+\alpha,\frac{1+\alpha}{2}\right)} \tag{1.10}$$

$$+ |b|_{S_t}^{\left(1+\alpha,\frac{1+\alpha}{2}\right)} + |B|_{S_t}^{(\alpha,\alpha/2)}$$

*where* $t \le T$ *and* $C(t)$ *is a non-decreasing (exponential) function of* $t$.

In the case of Dirichlet boundary condition $\underset{\sim}{w} = \underset{\sim}{a}$ an analogous theorem was established in the papers [S 1–3]. The case $\sigma = 0$ was considered in the papers [S 4,5] where the following theorem was proved.

**Theorem 1.1'** *Suppose that* $\Gamma \in C^{2+\alpha}$ *and that* $\underset{\sim}{f},\rho,\underset{\sim}{w}_0,\underset{\sim}{d},b$ *satisfy conditions i)–iii) with* $\underset{\sim}{u} = 0$. *Then problem (1.8) with* $\sigma = 0$ , $B = 0$ *has a unique solution with the same differentiability properties as in the theorem 1.1, and the solution satisfies (1.10) without* $|B|_{S_t}^{\alpha,\alpha/2}$ *in the right-hand side.*

Let us make some comments concerning the condition iii). The representation formulas (1.9′) may be considered as restrictions on the structure of functions $\rho$ and $\underset{\sim}{f}$. Since the Stokes system is not parabolic but rather degenerate parabolic, it can be shown that without any restrictions of this type at all, the solution looses regularity with respect to t in comparison with regularity of $\underset{\sim}{f}$ and $\rho$. The resctrictions written in the form (1.9) are by no means necessary but convenient sufficient conditions preventing the loss of regularity. As it is seen from the proof of theorem 1.1, they permit to carry out such standard operations as smooth coordinate transformations and multiplication of solutions by smooth functions, although these operations change the Stokes system in an essential way. These conditions work in important particular cases, for instance, they are satisfied in the case $\rho = 0$, $\nabla \cdot \underset{\sim}{f} = 0$, since we may set $\underset{\sim}{h} = 0$, $\underset{\sim}{h}_k = 0$ (hence, in this case the solutions has optimal regularity properties). On the other hand, if $\rho = 0$, but $\underset{\sim}{f}$ is an arbitrary vector fields from $C^{\alpha,\alpha/2}(Q_T)$, the condition iii) does not seem to hold, for if we take $\underset{\sim}{h} = \underset{\sim}{f}$ and represent $\underset{\sim}{f}$ in a divergence form

$$\underset{\sim}{f} = \sum_{k=1}^{3} \frac{\partial \underset{\sim}{h}_k}{\partial \xi_k}$$

with

$$\underset{\sim}{h}_k(\xi,t) = -\frac{1}{4\pi} \frac{\partial}{\partial \xi_k} \int_\Omega \underset{\sim}{f}(\eta,t) \frac{d\eta}{|\xi - \eta|}$$

then the conditions $|\underset{\sim}{h}_k|_{Q_T}^{(1+\alpha,\gamma)} < \infty$ require more regularity of $\underset{\sim}{f}$ with respect to t, namely,

$$\underset{\sim}{f} \in C^{\alpha,\alpha/2+\varepsilon/2}(Q_T) \text{ with } \varepsilon = 1 - \gamma .$$

Other better choices of $\underset{\sim}{h}$ and $\underset{\sim}{h}_k$ do not seem to be possible, so in this case our theorem gives a loss of regularity by $\varepsilon/2$.

For the problem (1.7) we obtain the following result.

**Theorem 1.2** *Suppose that* $\Gamma \in C^{2+\alpha}$, $\sigma \in C^\alpha(\Gamma)$, $\sigma \geq \sigma_0 > 0$ *and that*

$$\underset{\sim}{u} \in C^{2+\alpha,1+\alpha/2}(Q_T),$$

*satisfies the inequalities*

$$\left(T + T^{1/2}\right) |\underset{\sim}{u}|_{Q_T}^{(2+\alpha,1+\alpha/2)} \leq \delta \qquad (1.11_1)$$

$$T^{\frac{1-\alpha+\gamma}{2}} |D\underset{\sim}{u}|_{Q_T} \leq \delta \qquad (1.11_2)$$

*where*

$$D\underset{\sim}{u} = \left\{\frac{\partial u_i}{\partial x_k}\right\}_{i,k=1,2,3},$$

*and $\delta$ is a sufficiently small positive number (see condition (5.24) below). Assume also that conditions i)–iii) are satisfied. Then problem (1.7) has a unique solution with the same regularity properties as in theorem 1.1., and*

$$|\underset{\sim}{w}|_{Q_t}^{(2+\alpha,1+\alpha/2)} + |\nabla s|_{Q_t}^{(\alpha,\alpha/2)} + |s|_{S_t}^{\left(1+\alpha,\frac{1+\alpha}{2}\right)} + |s|_{Q_t}^{(1+\alpha,\gamma)}$$

$$\le C'(t)\Big[F(t) + P_{t_0}[\underset{\sim}{u}](|w_0|_\Omega + |Dw_0|_\Omega + |b(.,0)|_\Gamma)\Big] \qquad (1.12)$$

*where $C'(t)$ is a non-decreasing function of $t \le T$, $F(t)$ is the sum of norms in the right-hand side of (1.10), and*

$$P_t[\underset{\sim}{u}] = t^{\frac{1-\alpha}{2}}(|\underset{\sim}{u}|_{Q_t} + |D\underset{\sim}{u}|_{Q_t}) + |D\underset{\sim}{u}|_{Q_t}^{(\alpha,\alpha/2)} + \langle D\underset{\sim}{u}\rangle_{t,Q_t}^{\left(\frac{1+\alpha-\gamma}{2}\right)}. \qquad (1.13)$$

We present finally local existence theorem for nonlinear problem (1.6).

**Theorem 1.3** *Assume that $\Gamma \in C^{3+\alpha}$ and that*

$$\underset{\sim}{f},\ \underset{\sim}{f}_{x_k} \in C^{\alpha,\frac{\alpha+\varepsilon}{2}}(R^3 x(0,T_0)) \text{ with } \alpha \in (0,1),\ \varepsilon\in(0,1-\alpha).$$

*For arbitrary $\underset{\sim}{v}_0 \in C^{2+\alpha}(\Omega)$ satisfying the compatibility conditions*

$$\nabla\cdot\underset{\sim}{v}_0 = 0,\ \Pi_0 S(\underset{\sim}{v}_0)\underset{\sim}{n}_0\big|_\Gamma = 0$$

*problem (1.6) is uniquely solvable in a finite time interval $(0,T')$ whose magnitude $T' \le T_0$ depends on the data (see condition (6.17) below). The solution possesses the same regularity properties as in theorems 1.1, 1.2.*

At the conclusion we notice that the proof of theorems 1.2 and 1.3 given below is valid also in the case $\sigma = 0$ and provides a local existence theorem established earlier in [S 6]. The proof of estimate (1.10) for the solution of problem (1.8) which is given in §3 is independent of theorem 1.1′ but we have found it convenient to use this theorem in proving the solvability of problem (1.8).

## §2. Auxiliary constructions and inequalities.

In this section we present some auxiliary calculations and estimates which are necessary for the proof of theorem 1.1. We recall that the condition $\Gamma \in C^{2+\alpha}$ means that in the neighbourhood of every point $\xi_0 \in \Gamma$ the surface $\Gamma$ may be determined by the equation

$$y_3 = \varphi(y_1,y_2) \tag{2.1}$$

in a so called local cartesian coordinate system $(y_1,y_2,y_3)$ with the origine in $\xi_0$ and with the $y_3$ axis directed along the interior normal $-\underset{\sim}{n}_0(\xi_0)$. The function $\varphi$ is defined in a disc

$$K_d:\ |y'| \le d;$$

moreover,

$$\varphi \in C^{2+\alpha}(K_d)$$

and

$$|\varphi|^{(2+\alpha)}_{K_d} \le M, \tag{2.2}$$

the constants d and M being independent of $\xi_0$. Clearly,

$$\varphi(0) = 0,\ \nabla\varphi(0) = 0,$$

and

$$|\varphi(y')| \le |\varphi(y') - \varphi(0)| \le M|y'|,$$

$$|\nabla\varphi(y')| \le |\nabla\varphi(y') - \nabla\varphi(0)| \le M|y'|.$$

It is convenient to extend the function $\varphi$ from $K_d$ into the whole plane $\mathbb{R}^2$ with the preservation of class. We shall assume that this extension is already made and that the estimate (2.2) holds for $d = \infty$, i.e. in $\mathbb{R}^2$.

In the coordinate system $(y_1,y_2,y_3)$, the components of exterior normal vector $\underset{\sim}{n}_0(y)$ are equal to

$$n_{01} = \frac{\varphi_{y_1}}{\sqrt{1+|\nabla\varphi|^2}},\ n_{02} = \frac{\varphi_{y_2}}{\sqrt{1+|\nabla\varphi|^2}},\ n_{03} = -\frac{1}{\sqrt{1+|\nabla\varphi|^2}}. \tag{2.3}$$

The Laplace-Beltrami operator on $\Gamma$ is given by the formula

$$\Delta = \frac{1}{\sqrt{g}}\sum_{\alpha,\beta=1}^{2} \frac{\partial}{\partial S_\alpha} g^{\alpha\beta}\sqrt{g}\,\frac{\partial}{\partial S_\beta} \tag{2.4}$$

where $(S_1,S_2)$ are arbitrary local coordinates on $\Gamma$,

$$g = \det(g_{\alpha\beta})_{\alpha,\beta=1,2},\ g_{\alpha\beta} = \sum_{l=1}^{3} \frac{\partial\xi_l}{\partial S_\alpha}\cdot\frac{\partial\xi_l}{\partial S_\beta}$$

are elements of the metric tensor, and $g^{\alpha\beta}$ are elements of the inverse matrix to $(g_{\alpha\beta})$, i.e.

$$g^{11} = \frac{g_{22}}{g},\quad g^{22} = \frac{g_{11}}{g},\ g^{12} = g^{21} = -\frac{g_{12}}{g}.$$

If we take $(y_1,y_2)$ as local coordinates of the point$(y_1,y_2,\varphi(y_1,y_2)) \in \Gamma$, then

$$g_{\alpha\beta} = \sum_{l=1}^{3}\frac{\partial y_l}{\partial y_\alpha}\cdot\frac{\partial y_l}{\partial y_\beta} = \delta_{\alpha\beta} + \varphi_{y_\alpha}\varphi_{y_\beta},\ g = 1 + |\nabla\varphi|^2,$$

and

$$g^{\alpha\beta} = \delta_{\alpha\beta} - \frac{\varphi_{y_\alpha}\varphi_{y_\beta}}{1 + |\nabla\varphi|^2}.$$

Hence,

$$\Delta = \frac{1}{\sqrt{g}}\sum_{\alpha,\beta=1}^{2}\frac{\partial}{\partial y_\alpha}\ g^{\alpha\beta}\sqrt{g}\,\frac{\partial}{\partial y_\beta} = \Delta' + \Delta'' \tag{2.5}$$

where

$$\Delta' = \frac{\partial^2}{\partial y_1^2} + \frac{\partial^2}{\partial y_2^2}$$

is the Laplacian on the tangent plane to $\Gamma$ at the origine, and

$$\Delta'' = -\frac{1}{g}\sum_{\alpha,\beta=1}^{2} \varphi_{y_\alpha}\varphi_{y_\beta}\frac{\partial^2}{\partial y_\alpha \partial y_\beta} + \frac{1}{\sqrt{g}}\sum_{\alpha=1}^{2}\frac{\partial\sqrt{g}}{\partial y_\alpha}\frac{\partial}{\partial y_\alpha}$$

$$+ \sum_{\alpha,\beta=1}^{2}\frac{\partial}{\partial y_\beta}\left(\frac{\varphi_{y_\alpha}\varphi_{y_\beta}}{\sqrt{g}}\right)\frac{\partial}{\partial y_\alpha}. \tag{2.6}$$

Suppose that $y_i$ and $\xi_k$ are related to each other by a linear transformation $\mathfrak{C}_{\xi_0}$

$$y = \mathfrak{C}(\xi - \xi_0)$$

where $\mathfrak{C}$ is an orthogonal matrix, and let

$$z = Yy = (y_1, y_2, y_3 - \varphi(y')).$$

Then $\mathcal{Z}_{\xi_0} = \mathfrak{C}_{\xi_0}^{-1} Y \mathfrak{C}_{\xi_0}$ maps the domain $y_3 > \varphi(y')$ into the half-space

$$D(\xi_0)\ (\underset{\sim}{z} - \underset{\sim}{z}_0)\cdot \underset{\sim}{n}(\xi_0) < 0.$$

The Jacobian matrix $\mathcal{J}(\xi)$ of this transformation possesses the properties

$$\det \mathcal{J} = 1,\ \mathcal{J}(\xi_0) = \mathbb{I}.$$

Let $(\underset{\sim}{w}, s)$ be a solution of problem (1.8) and let $\zeta_\lambda(\xi)$, $\lambda \le \alpha/2$, be a $C_0^\infty(\mathbb{R}^3)$ - function such that

$$\zeta_\lambda(\xi) = 1 \text{ for } |\xi - \xi_0| \le \lambda,\ \zeta_\lambda(\xi) = 0 \text{ for } |\xi - \xi_0| \ge 2\lambda$$

and

$$|D^k \zeta_\lambda(\xi)| \le C(k)\lambda^{-|k|}.$$

Multiplying (1.8) by $\zeta_\lambda$ we see immediately that $\underset{\sim}{w}_\lambda = \zeta_\lambda \underset{\sim}{w}$ and $s_\lambda = s\zeta_\lambda$ satisfy the following equations, initial and boundary conditions:

$$\underset{\sim}{w}_{\lambda t} - \nu\nabla^2\underset{\sim}{w}_\lambda + \nabla s_\lambda = \underset{\sim}{f}\zeta_\lambda - 2\nu(\nabla\zeta_\lambda\cdot\nabla)\underset{\sim}{w} - \nu\underset{\sim}{w}\nabla^2\zeta_\lambda$$
$$+ s\nabla\zeta_\lambda \equiv \underset{\sim}{f}^{(1)},$$
$$\nabla\cdot\underset{\sim}{w}_\lambda = \rho\zeta_\lambda + \underset{\sim}{w}\cdot\nabla\zeta_\lambda = \rho^{(1)},$$
$$\underset{\sim}{w}_\lambda\big|_{t=0} = \underset{\sim}{w}_0\zeta_\lambda \equiv \underset{\sim}{w}_{0\lambda},$$
$$\nu\Pi_0 S(\underset{\sim}{w}_\lambda)\underset{\sim}{n}_0 = \Pi_0\underset{\sim}{d}\zeta_\lambda + \nu\Pi_0\left(\underset{\sim}{w}\frac{\partial\zeta_\lambda}{\partial n} + (\underset{\sim}{w}\cdot\underset{\sim}{n}_0)\nabla\zeta_\lambda\right) \equiv \Pi_0\underset{\sim}{d}^{(1)}, \tag{2.7}$$
$$\underset{\sim}{n}_0\cdot T(\underset{\sim}{w}_\lambda, s_\lambda)\underset{\sim}{n}_0 - \sigma(\xi)\underset{\sim}{n}_0\cdot\Delta\int_0^t \underset{\sim}{w}_\lambda d\tau = b\zeta_\lambda + \int_0^t B\zeta_\lambda d\tau$$
$$+ 2\nu(\underset{\sim}{n}_0\cdot\underset{\sim}{w})\frac{\partial\zeta_\lambda}{\partial n} - \sigma(\xi)\underset{\sim}{n}_0\cdot\int_0^t [\zeta_\lambda\Delta\underset{\sim}{w} - \Delta(\underset{\sim}{w}\zeta_\lambda)]d\tau$$
$$\equiv b^{(1)} + \int_0^t B^{(1)}d\tau \;;$$

here

$$b^{(1)} = b\zeta_\lambda + 2\nu(\underset{\sim}{w}\cdot\underset{\sim}{n}_0)\frac{\partial\zeta_\lambda}{\partial n}$$

and

$$B^{(1)} = B\zeta_\lambda - \sigma(\xi)\underset{\sim}{n}_0\cdot\,[\zeta_\lambda\Delta\underset{\sim}{w} - \Delta(\underset{\sim}{w}\zeta_\lambda)].$$

After the coordinate transformation $z = \mathcal{Z}_{\xi_0}x$ these equations take the form

$$\hat{\underset{\sim}{w}}_{\lambda t} - \nu\hat{\nabla}^2\hat{\underset{\sim}{w}}_\lambda + \hat{\nabla}\hat{s}_\lambda = \hat{\underset{\sim}{f}}^{(1)}, \quad \hat{\nabla}\cdot\hat{\underset{\sim}{w}}_\lambda = \hat{\rho}^{(1)},$$
$$\hat{\underset{\sim}{w}}_\lambda\big|_{t=0} = \hat{\underset{\sim}{w}}_{0\lambda}, \tag{2.8}$$
$$\nu\hat{\Pi}_0\hat{S}(\hat{\underset{\sim}{w}}_\lambda)\underset{\sim}{n}_0 = \Pi_0\hat{\underset{\sim}{d}}^{(1)},$$
$$\hat{\underset{\sim}{n}}_0\cdot\hat{T}(\hat{\underset{\sim}{w}}_\lambda, \hat{s}_\lambda)\underset{\sim}{n}_0 - \hat{\sigma}(z)\underset{\sim}{n}_0\cdot\hat{\Delta}\int_0^t \underset{\sim}{w}_\lambda d\tau = \hat{b}^{(1)} + \int_0^t \hat{B}^{(1)}d\tau$$

where $\hat{\underset{\sim}{w}}_\lambda(z,t) = \underset{\sim}{w}_\lambda(\mathcal{Z}_{\xi_0}^{-1}z,t)$ etc.,

$$\hat{\nabla} = \mathcal{J}^{*}\nabla = \left(\sum_{k=1}^{3} J_{km}\frac{\partial}{\partial z_k}\right)_{m=1,2,3}, \quad \hat{T}(\underset{\sim}{w},s) = -s\mathbb{I} + \hat{S}(\underset{\sim}{\hat{w}}_{\lambda}),$$

$$\hat{S}_{ij}(\underset{\sim}{w}) = \sum_{k=1}^{3}\left(J_{kj}\frac{\partial w_i}{\partial z_k} + J_{ki}\frac{\partial w_j}{\partial z_k}\right), \quad J_{km} = \frac{\partial z_k}{\partial \xi_m}$$

are elements of the Jacobi matrix $\mathcal{J}$.

Clearly, the first three equations in (2.8) hold for

$$z \in \mathbb{V}_d \equiv Z_{\xi_0}\Omega_d(\xi_0),$$

where

$$\Omega_d(\xi_0) = \{\xi \in \Omega: \ |\xi - \xi_0| < d\},$$

and the boundary conditions have a sense for $z \in \partial\mathbb{V}_d \cap \partial\mathbb{D}(\xi_0)$ but the operator $\hat{\nabla}$ and the functions $J_{mk}$ are defined in the whole half-space $\mathbb{D}(\xi_0)$ and $\underset{\sim}{\hat{w}}_{\lambda}(z,t) = 0$, $s_{\lambda}(z,t) = 0$ for $z \in \mathbb{V}_d\backslash\mathbb{V}_{2\lambda}$.

Therefore, if we extend $\hat{\sigma}(z)$ into the whole plane $\mathbb{R}(\xi_0) \equiv \partial\mathbb{D}(\xi_0)$ with the preservation of class and of the sign, and also extend $\underset{\sim}{w}_{\lambda}$, $s_{\lambda}$ by zero into $\mathbb{D}(\xi_0)\backslash\mathbb{V}_d$ (as well as all the functions in the right-hand sides of the equations (2.8)), then we may consider (2.8) as a half-space initial-boundary value problem for $(\underset{\sim}{\hat{w}}_{\lambda},\hat{s}_{\lambda})$. We rewrite this problem in the form

$$\underset{\sim}{\hat{w}}_{\lambda t} - \nu\hat{\nabla}^2\underset{\sim}{\hat{w}}_{\lambda} + \hat{\nabla}s_{\lambda} = \underset{\sim}{\hat{f}}^{(1)} + \underset{\sim}{k}_1(\underset{\sim}{\hat{w}}_{\lambda},\hat{s}_{\lambda}) \equiv \underset{\sim}{f}^{(2)}$$

$$\hat{\nabla}\cdot\underset{\sim}{w}_{\lambda} = \hat{\rho}^{(1)} + k_2(\underset{\sim}{\hat{w}}_{\lambda}) \equiv \rho^{(2)} \qquad (z \in \mathbb{D}(\xi)),$$

$$\underset{\sim}{\hat{w}}_{\lambda}\big|_{t=0} = \underset{\sim}{\hat{w}}_{0\lambda},$$

$$\nu\Pi_{00}\hat{S}(\underset{\sim}{w}_{\lambda})\underset{\sim}{n}_{00} = \Pi_{00}\hat{\Pi}_0\underset{\sim}{d}^{(1)} + \underset{\sim}{k}_3(\underset{\sim}{\hat{w}}_{\lambda}) \equiv \Pi_{00}\underset{\sim}{d}^{(2)} \tag{2.9}$$

$$\underset{\sim}{n}_{00}\cdot T(\underset{\sim}{w}_{\lambda},s_{\lambda})\underset{\sim}{n}_{00} - \sigma(\xi_0)\underset{\sim}{n}_{00}\cdot\Delta'\int_0^t\underset{\sim}{\hat{w}}_{\lambda}d\tau$$

$$= \hat{b}^{(1)} + \int_0^t\hat{B}^{(1)}d\tau + k_4(\underset{\sim}{\hat{w}}_{\lambda}) + \int_0^t k_5(\underset{\sim}{\hat{w}}_{\lambda})d\tau$$

$$\equiv b^{(2)} + \int_0^t B^{(2)}d\tau$$

where $\mathbf{n}_{00} = \mathbf{n}_0(\xi_0)$, $\Pi_{00}\mathbf{f} = \mathbf{f} - \mathbf{n}_{00}(\mathbf{f}\cdot\mathbf{n}_{00})$ is a projection of the vector $\mathbf{f}$ onto the plane $\mathbb{R}(\xi_0) = \partial\mathbb{D}(\xi_0)$, $\Delta'$ is the Laplacian on this plane,

$$b^{(2)} = \hat{b}^{(1)} + k_4(\hat{\mathbf{w}}_\lambda)\ ,\ B^{(2)} = \hat{B}^{(1)} + k_5(\hat{\mathbf{w}}_\lambda)$$

and

$$\mathbf{k}_1(\mathbf{w},s) = \nu(\hat{\nabla}^2 - \nabla^2)\mathbf{w} + (\nabla - \hat{\nabla})s,$$

$$k_2(\mathbf{w}) = (\nabla - \hat{\nabla})\cdot\mathbf{w},$$

$$\mathbf{k}_3(\mathbf{w}) = \nu\Pi_{00}\hat{\mathbf{n}}_0(\hat{\mathbf{n}}_0\cdot\hat{S}(\mathbf{w})\hat{\mathbf{n}}_0) + \nu\Pi_{00}(S(\mathbf{w})\mathbf{n}_{00} - \hat{S}(\mathbf{w})\hat{\mathbf{n}}_0),$$

$$k_4(\mathbf{w}) = \nu[\mathbf{n}_{00}\cdot S(\mathbf{w})\mathbf{n}_{00} - \hat{\mathbf{n}}_0\cdot\hat{S}(\mathbf{w})\hat{n}_0]\ ,$$

$$k_5(\mathbf{w}) = [\hat{\sigma}(z) - \hat{\sigma}(\xi_0)]\mathbf{n}_{00}\cdot\Delta'\mathbf{w} + \hat{\sigma}(z)(\hat{\mathbf{n}}_0\cdot\hat{\Delta}\mathbf{w} - \mathbf{n}_{00}\cdot\Delta'\mathbf{w})$$

From the property $\det\mathcal{J} = 1$ of the matrix $\mathcal{J}$ it follows that

$$\sum_{m=1}^{3} \frac{\partial}{\partial z_m} J_{mk} = 0 \qquad (k = 1,2,3)$$

(these equations can be verified directly if we consider $J_{mk}$ as cofactors of the elements $\dfrac{\partial\xi_k}{\partial z_m}$ in the matrix $\mathcal{J}^{-1}$).

As a consequence, $\hat{\nabla}\cdot\mathbf{u} = \nabla\cdot\mathcal{J}\mathbf{u}$, and the expressions $\mathbf{k}_1(\mathbf{w},s)$, $k_2(\mathbf{w})$ may be written in a divergence form as follows:

$$k_2(\mathbf{w}) = \nabla\cdot(I - \mathcal{J})\mathbf{w},\ \mathbf{k}_1(\mathbf{w},s) = \sum_{m=1}^{3} \frac{\partial \mathbf{k}_{1m}(\mathbf{w},s)}{\partial\xi_m},$$

$$\mathbf{k}_{1m} = \sum_{j,l=1}^{3} (J_{mj}J_{lj} - \delta_{ml})\frac{\partial\mathbf{w}}{\partial z_l} + \sum_{j=1}^{3} (\delta_{mj} - J_{mj})\mathbf{e}_j s\ , \tag{2.10}$$

$$\mathbf{e}_j = (\delta_{1j},\ \delta_{2j},\ \delta_{3j}).$$

For subsequent estimates, it is important to write convenient representation formulas for $\dfrac{\partial\rho^{(2)}}{\partial t} - \nabla\cdot\mathbf{f}^{(2)}$, as required in condition iii), §1. In deriving these formulas, we restrict ourselves with formal calculations which can be interpreted quite rigorously by means of appropriate integral identities. We assume that the functions $\rho$ and $\mathbf{f}$ in the problem (1.8) under consideration safisfy condition iii).

First of all, we have

$$\frac{\partial \rho^{(1)}}{\partial t} - \nabla\cdot \underset{\sim}{f}^{(1)} = \zeta_\lambda\left(\frac{\partial \rho}{\partial t} - \nabla\cdot\underset{\sim}{f}\right) + \nabla\zeta_\lambda(\underset{\sim}{w}_t - \underset{\sim}{f})$$

$$+ \nabla\cdot[2\dot{\nu}(\nabla\zeta_\lambda\cdot\nabla)\underset{\sim}{w} + \nu\underset{\sim}{w}\,\nabla^2\zeta_\lambda - s\nabla\zeta_\lambda].$$

If we make use of the equation $\underset{\sim}{w}_t - \underset{\sim}{f} = \nu\nabla^2\underset{\sim}{w} - \nabla s$ and apply well known formulas of vector analysis

$$\nabla^2\underset{\sim}{w} = - \text{rot rot}\underset{\sim}{w} + \nabla(\nabla\cdot\underset{\sim}{w}),\ \nabla\cdot[\underset{\sim}{\varphi}\times\underset{\sim}{\psi}] = \underset{\sim}{\psi}\ \text{rot}\underset{\sim}{\varphi} - \underset{\sim}{\varphi}\ \text{rot}\underset{\sim}{\psi},$$

we easily show that

$$\frac{\partial \rho^{(1)}}{\partial t} - \nabla\cdot\underset{\sim}{f}^{(1)} = \nabla\cdot\underset{\sim}{h}^{(1)} + j$$

with

$$\underset{\sim}{h}^{(1)} = \zeta_\lambda\underset{\sim}{h} + \underset{\sim}{h}',$$

$$\underset{\sim}{h}' = 2\nu(\nabla\zeta_\lambda\cdot\nabla)\underset{\sim}{w} + \nu\underset{\sim}{w}\nabla^2\zeta_\lambda - s\nabla\zeta_\lambda + \nu[\nabla\zeta_\lambda\times\text{rot }\underset{\sim}{w}] \tag{2.11}$$

$$- (s - \nu\nabla\cdot\underset{\sim}{w})\nabla\zeta_\lambda,$$

$$j = -\nabla\zeta_\lambda\cdot\underset{\sim}{h} + (s - \nu\nabla\cdot\underset{\sim}{w})\nabla^2\zeta_\lambda. \tag{2.12}$$

Hence,

$$\frac{\partial \rho^{(2)}}{\partial t} - \nabla\cdot\underset{\sim}{f}^{(2)} = \frac{\partial \hat{\rho}^{(1)}}{\partial t} - \nabla\cdot\hat{\underset{\sim}{f}}^{(1)} + (\hat{\nabla} - \hat{\nabla})\cdot\underset{\sim}{w}_{\lambda t} - \nabla\cdot\underset{\sim}{k}_1$$

$$= \hat{\nabla}\cdot\hat{\underset{\sim}{h}}^{(1)} + \hat{j} + (\hat{\nabla} - \hat{\nabla})(w_{\lambda t} - \hat{\underset{\sim}{f}}^{(1)}) - \nabla\cdot\underset{\sim}{k}_1.$$

Writing $\hat{j}$ in the form

$$\hat{j}(z,t) = -\frac{1}{4\pi}\nabla^2\int_{D(\xi_0)}\frac{\hat{j}(y,t)}{|z-y|}\,dy$$

we obtain

$$\frac{\partial \rho^{(2)}}{\partial t} - \nabla\cdot \underset{\sim}{f}^{(2)} = \nabla\cdot \underset{\sim}{h}^{(2)},$$

$$\hat{\underset{\sim}{h}}^{(2)} = \mathcal{J}\hat{\underset{\sim}{h}}^{(1)} + (I - \mathcal{J})(\hat{\underset{\sim}{w}}_{\lambda t} - \hat{\underset{\sim}{f}}^{(1)}) - \underset{\sim}{k}_1$$

$$- \frac{1}{4\pi}\nabla\int_{D(\xi_0)} \frac{\hat{j}(y,t)}{|z - y|}\, dy = \mathcal{J}\hat{\underset{\sim}{h}}^{(1)} + (I - \mathcal{J})(\nu\nabla^2\hat{\underset{\sim}{w}}_{\lambda t} - \nabla\hat{s}_\lambda) \tag{2.13}$$

$$- \mathcal{J}\underset{\sim}{k}_1(\underset{\sim}{w}_\lambda, s_\lambda) - \frac{1}{4\pi}\nabla\int_{D(\xi_0)} \frac{\hat{j}(y,t)}{|z - y|}\, dy.$$

It remains to represent $\underset{\sim}{h}^{(2)}$ in divergence form. Condition iii) for $\underset{\sim}{h}$ implies

$$\underset{\sim}{h} = \sum_{m,k=1}^{3} \frac{\partial}{\partial z_m} J_{mk}\hat{\underset{\sim}{h}}_k,$$

consequently,

$$\hat{\underset{\sim}{h}}^{(1)} = \sum_{m,k=1}^{3} \left(\frac{\partial}{\partial z_m} \hat{\zeta}_\lambda J_{mk}\hat{\underset{\sim}{h}}_k - \frac{\partial\hat{\zeta}_\lambda}{\partial z_m} J_{mk}\hat{\underset{\sim}{h}}_k + \hat{\underset{\sim}{h}}'\right)$$

and

$$\hat{\underset{\sim}{h}}^{(2)}(z,t) = \sum_{m=1}^{3} \frac{\partial \underset{\sim}{h}_m^{(2)}}{\partial z_m}$$

with

$$\hat{\underset{\sim}{h}}_m^{(2)} = \sum_{k=1}^{3} J_{mk}\hat{\zeta}_\lambda\mathcal{J}\hat{\underset{\sim}{h}}_k + (I - \mathcal{J})(\nu\frac{\partial\hat{\underset{\sim}{w}}_\lambda}{\partial z_m} - \underset{\sim}{e}_m\hat{s}_\lambda) - \mathcal{J}\underset{\sim}{k}_{1m}$$

$$- \frac{1}{4\pi}\frac{\partial}{\partial z_m}\int_{D(\xi_0)} \frac{\underset{\sim}{h}^{(3)}(y,t)}{|z - y|}\, dy - \frac{1}{4\pi}\underset{\sim}{e}_m\int_{D(\xi_0)} \frac{\hat{j}(y,t)}{|z - y|}\, dy, \tag{2.14}$$

$$\hat{\underset{\sim}{h}}^{(3)} = \mathcal{J}\hat{\underset{\sim}{h}}' - \sum_{l,k=1}^{3} \left(\frac{\partial\zeta_\lambda}{\partial z_l} J_{lk}\mathcal{J}\hat{\underset{\sim}{h}}_k + \zeta_\lambda J_{lk}\frac{\partial\mathcal{J}}{\partial z_l}\hat{\underset{\sim}{h}}_k\right)$$

$$+ \sum_{l=1}^{3} \frac{\partial\mathcal{J}}{\partial z_l}\left(\nu\frac{\partial\underset{\sim}{w}_\lambda}{\partial z_l} - \underset{\sim}{e}_l s_\lambda + \underset{\sim}{k}_{1l}\right).$$

Consider also the case when the ball $|\xi - \xi_0| < 2\lambda$ is contained in $\Omega$. Then $(\tilde{w}_\lambda, s_\lambda)$ is a solution of the Cauchy problem

$$\begin{aligned} \tilde{w}_{\lambda t} - \nu\nabla^2\tilde{w}_\lambda + \nabla s_\lambda &= \tilde{f}\zeta_\lambda - 2\nu(\nabla\zeta_\lambda\cdot\nabla)\tilde{w} \\ &- \nu\tilde{w}\nabla^2\zeta_\lambda + s\nabla\zeta_\lambda = \tilde{f}^{(1)}, \end{aligned} \tag{2.15}$$

$$\nabla\cdot\tilde{w}_\lambda = \nabla\zeta_\lambda\cdot\tilde{w} + \rho\zeta_\lambda = \rho^{(1)}, \qquad \tilde{w}_\lambda|_{t=0} = \tilde{w}_0\zeta_\lambda = \tilde{w}_{0\lambda},$$

and

$$\frac{\partial\rho^{(1)}}{\partial t} - \nabla\cdot\tilde{f}^{(1)} = \nabla\cdot\tilde{h}^{(1)} + j = \nabla\cdot\tilde{H}^{(1)},$$

$$\tilde{H}^{(1)} = \tilde{h}^{(1)} - \frac{1}{4\pi}\nabla\int_{\mathbb{R}^3}\frac{j(y,t)}{|\xi - y|}\,dy = \sum_{m=1}^{3}\frac{\partial \tilde{H}_m^{(1)}}{\partial\xi_m},$$

$$\tilde{H}_m^{(1)} = \zeta_\lambda\tilde{h}_m - \frac{1}{4\pi}\tilde{e}_m\int_{\mathbb{R}^3}\frac{j(y,t)}{|\xi - y|}\,dy$$

$$+ \frac{1}{4\pi}\frac{\partial}{\partial\xi_m}\int_{\mathbb{R}^3}\left(\sum_{l=1}^{3}\frac{\partial\zeta_\lambda}{\partial y}\tilde{h}_l - \tilde{h}'(y,t)\right)\frac{dy}{|\xi - y|}.$$

Let us now present some auxiliary inequalities. We recall that for arbitrary function $u \in C^l(G)$, $G \in \mathbb{R}^n$, the following "interpolation inequalities" hold (see for instance [ADN]):

$$\langle u\rangle^{(l_1)} \le \varepsilon^{l-l_1}\langle u\rangle_G^{(l)} + C_1\varepsilon^{-l_1}|u|_G, \tag{2.16$_1$}$$

$$|D^k u|_G \le \varepsilon^{l-|k|}\langle u\rangle_G^{(l)} + C_2\varepsilon^{-|k|}|u|_G. \tag{2.16$_2$}$$

Here $l_1<l$, $|k|<l$, $\varepsilon$ is an arbitrary small constant (say, $\varepsilon \in (0,1)$) and $C_1, C_2$ are independent of $\varepsilon$. Now, for arbitrary $u,v \in C^l(G)$ we have

$$|uv|_G^{(l)} \le C_3|u|_G^{(l)}|v|_G^{(l)}, \tag{2.17}$$

moreover, the following (in some respect more precise) inequality holds true:

$$|uv|_G^{(l)} \le (|u|_G + \varepsilon|u|_G^{(l)})\langle v\rangle_G^{(l)} + C_4(\varepsilon)|u|_G^{(l)}|v|_G \tag{2.18}$$

For $l \in (0,1)$ this inequality is obvious, since in this case

$$\langle uv\rangle_G^{(l)} \le |u|_G \langle v\rangle_G^{(l)} + |v|_G \langle u\rangle_G^{(l)} \tag{2.19}$$

In the case $l > 1$ it can be obtained in the following way: we estimate the norms $\langle D^k uv\rangle_G^{(l-[l])}$, $|k|=[l]$, using (2.19) and then apply interpolation inequalities (2.16) to estimate $|D^i v|_G$, $|i| \le [l]$ and $\langle D^j v\rangle_G^{(l-[l])}$ with $|j| < l$.

Similar inequalities hold for functions depending on $(x,t) \in G \times (0,T) = \mathcal{G}_T$. If $l_1 \le l$ and $2|k| + a < l$, then

$$\langle u\rangle_{\mathcal{G}_T}^{(l_1, l_1/2)} \le \varepsilon^{l-l_1} \langle u\rangle_{\mathcal{G}_T}^{(l, l/2)} + C_5 \varepsilon^{-l_1} |u|_{\mathcal{G}_T}, \tag{$2.20_1$}$$

$$|D_x^k D_t^a u|_{\mathcal{G}_T} = \varepsilon^{(l-2|k|-a)} \langle u\rangle_{\mathcal{G}_T}^{(l,l/2)} + C_6 \varepsilon^{-(2|k|+a)} |u|_{\mathcal{G}_T} \tag{$2.20_2$}$$

with arbitrary $\varepsilon \in (0,1)$. Unfortunately, the constants $C_5$ and $C_6$ are in general unbounded as $T \to 0$ but this does not occur if the function $u(x,t)$ vanish for $t = 0$ together with all derivatives $D_t^a u$, $a < l/2$, since in this case we can extend $u(x,t)$ by zero into the half-space $t < 0$ and prove (2.20) for extended functions.

The following lemma will be used in §4.

**Lemma 2.1** *Any* $v \in C^{2+\alpha, 1+\alpha/2}(\mathcal{G}_T)$ *satisfies the inequality*

$$|v_t|_{\mathcal{G}_T} \le \varepsilon^{\alpha} \langle v_t\rangle_{t,\mathcal{G}_T}^{(\alpha/2)} + C_7 \varepsilon^{-1} (|v|_{\mathcal{G}_T} + |v_t(\cdot,0)|_G) \tag{2.21}$$

*where* $\varepsilon \in (0,1)$ *is arbitrary and* $C_7$ *is a nondecreasing function of* $T$.

Proof. (2.21) is an immediate consequence of a one-dimensional inequality

$$|u_t|_{\mathcal{G}_T} \le \varepsilon^{\alpha} \langle u_t\rangle_{t,\mathcal{G}_T}^{(\alpha/2)} + C_8 \varepsilon^{-1} |u|_{\mathcal{G}_T}$$

where $u(\xi,t) = v(\xi,t) - t v_t(\xi,0) - v(\xi,0)$ for $t > 0$ and $u(\xi,t) = 0$ for $t < 0$.

**Lemma 2.2** *If* $u \in C^{1+\alpha,\frac{1+\alpha}{2}}(\mathcal{G}_T)$, *then*

$$\langle u\rangle^{(1+\alpha,\gamma)}_{\mathcal{G}_T} \le C_9\varepsilon^{1-\gamma}\langle\nabla u\rangle^{(\alpha/2)}_{t,\mathcal{G}_T} + 2\varepsilon^{-\gamma}\langle u\rangle^{\left(\frac{1+\alpha}{2}\right)}_{t,\mathcal{G}_T}, \ \forall\varepsilon \in (0,1).$$

Proof. Consider the expression

$$V = |x - y|^{-\gamma}|t - t'|^{-\frac{1+\alpha-\gamma}{2}}|u(x,t)-u(y,t)-u(x,t')+u(y,t')|.$$

If $|t - t'| \ge \varepsilon^{-2}|x - y|^2$, then

$$V \le \varepsilon^{1-\gamma}|x - y|^{-1}|t - t'|^{-\alpha/2}|u(x,t)| - u(y,t) - u(x,t')$$

$$+ u(y,t')| \le C_9\varepsilon^{1-\gamma}\sup_{z\in G}\frac{|\nabla u(z,t) - \nabla u(z,t')|}{|t - t'|^{\alpha/2}} \le C_9\varepsilon^{1-\gamma}\langle\nabla u\rangle^{(\alpha/2)}_{t,\mathcal{G}_T}.$$

In the case $|t - t'| \le \varepsilon^{-2}|x - y|^2$ we have

$$V \le \varepsilon^{-\gamma}\left(\frac{|u(x,t) - u(x,t')|}{|t - t'|^{\frac{1+\alpha}{2}}} + \frac{|u(y,t) - u(y,t')|}{|t - t'|^{\frac{1+\alpha}{2}}}\right) \le 2\varepsilon^{-\gamma}\langle u\rangle^{\left(\frac{1+\alpha}{2}\right)}_{t,\mathcal{G}_T}.$$

Hence

$$\langle u\rangle^{(1+\alpha,\gamma)}_{\mathcal{G}_T} \le \sup_{x,y,t,t'} V \le C_9\varepsilon^{1-\gamma}\langle\nabla u\rangle^{(\alpha/2)}_{t,\mathcal{G}_T} + 2\varepsilon^{-\gamma}\langle u\rangle^{\left(\frac{1+\alpha}{2}\right)}_{t,\mathcal{G}_T},$$

and lemma is proved.

As a corollary, we easily obtain the estimate

$$|u|^{(1+\alpha,\gamma)}_{\mathcal{G}_T} \le C_{10}|u|^{\left(1+\alpha,\frac{1+\alpha}{2}\right)}_{\mathcal{G}_T}.$$

Further, arbitrary $u,v \in C^{1,1/2}(\mathcal{G}_T)$ satisfy the inequality

$$|uv|^{(1,1/2)}_{\mathcal{G}_T} \le C_{11}|u|^{(1,1/2)}_{\mathcal{G}_T}|v|^{(1,1/2)}_{\mathcal{G}_T} \tag{2.22}$$

with a constant $C_{11}$ independent of T. Finally from the identity

$$u(x,t)v(x,t) - u(x,t-\tau)v(x,t-\tau)$$

$$= [u(x,t) - u(x,t-\tau)]v(x,t) + u(x,t-\tau)[v(x,t) - v(x,t-\tau)]$$

it follows that

$$|uv|_{\mathscr{G}_T}^{(1+\alpha,\gamma)} \le |u|_{\mathscr{G}_T}^{(1+\alpha,\gamma)}|v|_{\mathscr{G}_T} + |v|^{(1+\alpha,\gamma)}|u|_{\mathscr{G}_T}$$

$$+ \langle u\rangle_{t,\mathscr{G}_T}^{\left(\frac{1+\alpha-\gamma}{2}\right)} \sup_{t\le T} \langle v\rangle_G^{(\gamma)} + \langle v\rangle_{t,\mathscr{G}_T}^{\left(\frac{1+\alpha-\gamma}{2}\right)} \sup \langle u\rangle_G^{(\gamma)} \tag{2.23}$$

$$\le 2|u|_{\mathscr{G}_T}^{(1+\alpha,\gamma)} \sup_{t\le T} |v|_{G_T}^{(\gamma)} + 2|v|_{\mathscr{G}_T}^{(1+\alpha,\gamma)} \sup_{t\le T} |u|_G^{(\gamma)}$$

We may apply these estimates to first order and second order differential operators of the form

$$L_1u = \sum_{i=1}^{n} l_i'(\xi,t)u_{\xi_i},$$

$$L_2u = \sum_{i,j} l_{i,j}(\xi,t)u_{\xi_i\xi_j} + \sum_i l_i(\xi,t)u_{\xi_i}.$$

**Lemma 2.3.** *Suppose that* $l_{ij}, l_i \in C^{\alpha,\alpha/2}(\mathscr{G}_T)$ *and*

$$l_i' \in C^{1+\alpha,\frac{1+\alpha}{2}}(\mathscr{G}_T).$$

*Then*

$$|L_2u|_{\mathscr{G}_T}^{(\alpha,\alpha/2)} \le (C_{12}\max_{i,j} |e_{ij}|_{\mathscr{G}_T} + \varepsilon K)\langle u\rangle_{\mathscr{G}_T}^{(2+\alpha,1+\alpha/2)}$$
$$+ C_{13}(\varepsilon)K|u|_{\mathscr{G}_T}, \tag{2.24}$$

$$|L_1u|_{\mathscr{G}_T}^{\left(1+\alpha,\frac{1+\alpha}{2}\right)} \le (C_{14}\max_i |e_i'|_{\mathscr{G}_T} + \varepsilon K')\langle u\rangle_{\mathscr{G}_T}^{(2+\alpha,1+\alpha/2)}$$
$$+ C_{15}(\varepsilon)K'|u|_{\mathscr{G}_T} \tag{2.25}$$

*and, as a consequence,*

$$|L_2u|^{(\alpha,\alpha/2)}_{\mathcal{G}_T} \le C_{16}K|u|^{(2+\alpha,1+\alpha/2)}_{\mathcal{G}_T}, \tag{2.26}$$

$$|L_1u|^{\left(1+\alpha,\frac{1+\alpha}{2}\right)}_{\mathcal{G}_T} \le C_{17}K'\,|u|^{(2+\alpha,1+\alpha/2)}_{\mathcal{G}_T} \tag{2.27}$$

*where*

$$K = \max_{i,j}\,|l_{ij}|^{(\alpha,\alpha/2)}_{\mathcal{G}_T} + \max_i\,|l_i|^{(\alpha,\alpha/2)}_{\mathcal{G}_T},\ K' = \max_i\,|l'_i|^{\left(1+\alpha,\frac{1+\alpha}{2}\right)}_{\mathcal{G}_T}$$

<u>Proof.</u> Consider the principal part of $L_2$. We have

$$\langle l_{ij}u_{\xi_i\xi_j}\rangle^{(\alpha,\alpha/2)}_{\mathcal{G}_T} \le |l_{ij}|_{\mathcal{G}_T}\langle u_{\xi_i\xi_j}\rangle^{(\alpha,\alpha/2)}_{\mathcal{G}_T} + \langle l_{ij}\rangle^{(\alpha,\alpha/2)}_{\mathcal{G}_T}|u_{\xi_i\xi_j}|_{\mathcal{G}_T}.$$

Now, in virtue of $(2.16_2)$

$$|u_{\xi_i\xi_j}|_G \le \varepsilon^{\alpha}\langle u\rangle^{(2+\alpha)}_G + C_{18}(\varepsilon)|u|_G,\quad \forall\varepsilon\in(0,1) \tag{2.28}$$

which gives

$$\langle l_{ij}u_{\xi_i\xi_j}\rangle^{(\alpha,\alpha/2)}_{\mathcal{G}_T} \le \left(|l_{ij}|_{\mathcal{G}_T} + \varepsilon^{\alpha}\langle l_{ij}\rangle^{(\alpha,\alpha/2)}_{\mathcal{G}_T}\right)\langle u\rangle^{(2+\alpha,1+\alpha/2)}_{\mathcal{G}_T}$$

$$+ C_{18}(\varepsilon)\langle l_{ij}\rangle^{(\alpha,\alpha/2)}_{\mathcal{G}_T}|u|_{\mathcal{G}_T}.$$

Other terms, as well as $L_1u$, are estimated in a similar way.

At the conclusion, we give elementary estimates of the Newtonian potential

$$u(x,t) = \int_G \frac{f(y,t)}{|x-y|}\,dy.$$

In our application of these estimates, G will be a bounded domain $\Omega$, the whole space $\mathbb{R}^3$ or the half-space. In two latter cases $f(y,t)$ will have a compact support.

**Lemma 2.4** *Suppose that* $\operatorname{supp} f \subset K_r(x_0) = \{|x-x_0| \le r\}$ *where* $x_0 \in G$ *and* $r \le r_0$. *Then*

$$|u|^{(\alpha,\alpha/2)}_{\mathcal{G}_T} \le C_{19}(r_0)r^{2-\alpha}(|f|_{\mathcal{G}_T} + \langle f\rangle^{(\alpha/2)}_{t,\mathcal{G}_T}),$$

$$|\nabla u|^{(\alpha,\alpha/2)}_{\mathcal{G}_T} \le C_{20}r^{1-\alpha}(|f|_{\mathcal{G}_T} + \langle f\rangle^{(\alpha/2)}_{t,\mathcal{G}_T}),$$

$$|u|^{(1+\alpha,\gamma)}_{\mathcal{G}_T} + r|\nabla u|^{(1+\alpha,\gamma)}_{\mathcal{G}_T} \le C_{21}r^{2-\gamma}\langle f\rangle^{\left(\frac{1+\alpha-\gamma}{2}\right)}_{t,\mathcal{G}_T}.$$

All the inequalities follow from obvious estimates

$$|u(\cdot,t)|_G + r|\nabla u(\cdot,t)|_G + r^{\alpha}|u(\cdot,t)|^{(\alpha)}_G$$
$$+ r^{1+\alpha}|\nabla u(\cdot,t)|^{(\alpha)}_G \le C_{22}r^2|f(\cdot,t)|_G$$

applied either to $u(x,t)$, or to the difference $u(x,t) - u(x,t')$.

## §3. Proof of estimate (1.10).

We prove a-priori estimate (1.10) by well known J.Schauder's method which is based on preliminary consideration of the Cauchy problem

$$\begin{aligned} &\underset{\sim}{u}_t - \nu\nabla^2\underset{\sim}{u} + \nabla q = \underset{\sim}{f}, \quad \nabla\cdot\underset{\sim}{u} = \rho \qquad (\xi \in \mathbb{R}^3,\ 0 \le t \le T) \\ &\underset{\sim}{u}|_{t=0} = \underset{\sim}{u}_0 \end{aligned} \tag{3.1}$$

and of initial boundary-value problem (1.8) in a half-space (say, in $\mathbb{R}^3_+$: $x_3 > 0$). The following result is established in the paper [MS1].

**Theorem 3.1** *Let the data of half-space problem (1.8) with a constant $\sigma > 0$ satisfy the conditions i)–iii) of §1 with* $\underset{\sim}{u} = 0$. *Then this problem has a solution*

$$\underset{\sim}{w} \in C^{2+\alpha,1+\alpha/2}(D_T), \quad \nabla s \in C^{\alpha,\alpha/2}(D_T)$$

*satisfying the estimate*

$$\langle \underset{\sim}{w}\rangle^{(2+\alpha,1+\alpha/2)}_{\mathbb{D}_T} + \langle \nabla s\rangle^{(\alpha,\alpha/2)}_{\mathbb{D}_T} \le C_1(T)\Big\{\langle \underset{\sim}{f}\rangle^{(\alpha,\alpha/2)}_{\mathbb{D}_T}$$

$$+ \langle \underset{\sim}{w}_0\rangle^{(2+\alpha)}_{\mathbb{R}^3_+} + \langle \nabla\rho\rangle^{(\alpha,\alpha/2)}_{\mathbb{D}_T} + \langle \rho\rangle^{\left(1+\alpha,\frac{1+\alpha}{2}\right)}_{\mathbb{R}_T} + \langle \underset{\sim}{h}\rangle^{(\alpha,\alpha/2)}_{\mathbb{D}_T} \tag{3.2}$$

$$+ \sum_{k=1}^{3} \langle \underset{\sim}{h}_k\rangle^{(1+\alpha,\gamma)}_{\mathbb{D}_T} + \langle \underset{\sim}{d}\rangle^{\left(1+\alpha,\frac{1+\alpha}{2}\right)}_{\mathbb{R}_T} + \langle b\rangle^{\left(1+\alpha,\frac{1+\alpha}{2}\right)}_{\mathbb{R}_T} + \langle B\rangle^{(\alpha,\alpha/2)}_{\mathbb{R}_T}\Big\}.$$

*Here* $\mathbb{D}_T = \mathbb{R}^3_+ \times (0,T)$, $\mathbb{R}_T = \mathbb{R}^2 \times (0,T)$, $\mathbb{R}^2 = \partial\mathbb{R}^3_+$, $C_1(T)$ *is a nondecreasing (exponential) function of T, and*

$$\langle \underset{\sim}{h}_k\rangle^{(1+\alpha,\gamma)}_{\mathbb{D}_T} = \sup_{0<\tau<t<T}\ \sup_{x,y\in\mathbb{R}^3_+} \tau^{-\frac{1+\alpha-\gamma}{2}} |x-y|^{-\gamma} |\underset{\sim}{h}_k(x,t)$$

$$- \underset{\sim}{h}_k(x,t-\tau) - \underset{\sim}{h}_k(y,t) + \underset{\sim}{h}_k(y,t-\tau)|.$$

*The solution is unique in the class of bounded* $\underset{\sim}{w}, s$ *such that*

$$\underset{\sim}{w} \in C^{2+\alpha,1+\alpha/2}(\mathbb{D}_T),\ \nabla s \in C^{\alpha,\alpha/2}(\mathbb{D}_T).$$

Strictly speaking, the estimate of the solution in [MS1] slightly differs from (2.2), because there the sum of norms

$$\langle \nabla\rho\rangle^{(\alpha,\alpha/2)}_{\mathbb{D}_T} + \langle \rho\rangle^{\left(1+\alpha,\frac{1+\alpha}{2}\right)}_{\mathbb{R}_T}$$

is replaced by a stronger norm

$$\langle \rho\rangle^{\left(1+\alpha,\frac{1+\alpha}{2}\right)}_{\mathbb{D}_T}$$

(with a reference to the paper [S4]). However, a closer consideration of the arguments in this paper shows that this replacement is not necessary.

**Theorem 3.2** *If the functions* $\underset{\sim}{f}, \underset{\sim}{u}_0, \rho$ *satisfy the condition i)-iii) with* $\underset{\sim}{u} = 0$, *then the Cauchy problem (3.1) has a solution*

$$\underset{\sim}{u} \in C^{2+\alpha,1+\alpha/2}(\Pi_T), \quad \nabla\rho \in C^{\alpha,\alpha/2}(\Pi_T)$$

*satisfying the inequality*

$$\langle \underset{\sim}{u}\rangle^{(2+\alpha,1+\alpha/2)}_{\Pi_T} + \langle \nabla q\rangle^{(\alpha,\alpha/2)}_{\Pi_T} \le C_2(T)\Big\{\langle \underset{\sim}{f}\rangle^{(\alpha,\alpha/2)}_{\Pi_T} + \langle \underset{\sim}{u}_0\rangle^{(2+\alpha)}_{R^3}$$

$$+ \langle \nabla\rho\rangle^{(\alpha,\alpha/2)}_{\Pi_T} + \langle \underset{\sim}{h}\rangle^{(\alpha,\alpha/2)}_{\Pi_T} + \sum_{k=1}^{3} \langle h_k\rangle^{(1+\alpha,\gamma)}_{\Pi_T}\Big\}$$

*where* $\Pi_T = R^3 \times (0,T)$ *and* $C_2(T)$ *is a non-decreasing function of T. The solution is unique in the class of vector fields*

$$\underset{\sim}{u} \in C^{(2+\alpha,1+\alpha/2)}(\Pi_T),\ \nabla q \in C^{(\alpha,\alpha/2)}(\Pi_T)$$

*with a bounded q.*

We now proceed to the proof of inequality (1.10) for a solution of the problem (1.8) and start with the estimate of the sum

$$\langle \underset{\sim}{w}\rangle^{(2+\alpha,1+\alpha/2)}_{Q_T} + \langle \nabla s\rangle^{(\alpha,\alpha/2)}_{Q_T}.$$

It is easily seen that

$$\langle \underset{\sim}{w}\rangle^{(2+\alpha,1+\alpha/2)}_{Q_T} + \langle \nabla s\rangle^{(\alpha,\alpha/2)}_{Q_T} \le \sup_{x_0\in\Omega} \Big(\langle \underset{\sim}{w}\rangle^{(2+\alpha,1+\alpha/2)}_{Q^{(\lambda/4)}(x_0)}$$

$$+ \langle \underset{\sim}{\nabla s}\rangle^{(\alpha,\alpha/2)}_{Q^{(\lambda/4)}(x_0)}\Big) + 2\Big(\frac{4}{\lambda}\Big)^{\alpha}\Big(\sum_{|k|=2} |D^k_x \underset{\sim}{w}|_{Q_T} + |\underset{\sim}{w}_t|_{Q_T} + |\nabla s|_{Q_T}\Big) \tag{3.3}$$

for we can apply the estimate

$$\frac{|u(x,t) - u(y,t)|}{|x - y|^{\alpha}} \le \Big(\frac{4}{\lambda}\Big)^{\alpha} 2|u|_{Q_T}$$

to the functions $\dfrac{\partial^2 \underset{\sim}{w}}{\partial x_k \partial x_n}$, $\dfrac{\partial \underset{\sim}{w}}{\partial t}$ or to $\nabla s$ in the case $|x - y| \ge \lambda/4$.

If $\mathrm{dist}\,(x_0,\Gamma) \ge \lambda/2$, then we can estimate $(\underset{\sim}{w},s)$ as a solution of the Cauchy problem (2.15) (with $\lambda/4$ instead of $\lambda$). In the case $\mathrm{dist}(x_0,\Gamma) \le \lambda/2$ there exists a point $\xi_0 \in \Gamma$ such that $Q^{(\lambda/2)}(x_0) \subset Q^{(\lambda)}(\xi_0)$ and we obtain our estimate applying theorem 3.1 to problem (2.9).

We consider a more complicated second case. We "flatten" the boundary $\Gamma$ near the point $\xi_0$ and regard $(\underset{\sim}{w}\zeta_\lambda, s\zeta_\lambda)$ as a solution of problem (2.9). Before applying theorem 3.1, we estimate the expressions $k_1$. In virtue of lemma 2.3,

$$|\underset{\sim}{k}_1(\hat{\underset{\sim}{w}}_\lambda, \hat{s}_\lambda)|^{(\alpha,\alpha/2)}_{\mathbb{D}_T(\xi_0)} \le (K_1 + \varepsilon)\,(\langle \hat{\underset{\sim}{w}}_\lambda \rangle^{(2+\alpha,1+\alpha/2)}_{\mathbb{D}_T(\xi_0)}$$

$$+ \langle \nabla \hat{s}_\lambda \rangle^{(\alpha,\alpha/2)}_{\mathbb{D}_T(\xi_0)}) + C_3(\varepsilon) K_2 \left( |\hat{\underset{\sim}{w}}_\lambda|_{\mathbb{D}_T(\xi_0)} + |\nabla \hat{s}_\lambda|_{\mathbb{D}_T(\xi_0)} \right),$$

$$\mathbb{D}_T(\xi_0) = \mathbb{D}(\xi_0) \times (0,T),$$

where $K_1$ is the sum of $L_\infty$-norms of the leading coefficients of $\underset{\sim}{k}_1$ in a ball $|z| \le 2\lambda$, and $K_2$ is the sum of $C^\alpha$-norms of all the coefficients (they do not depend on t). Since all the leading coefficients contain factors $J_{mk} - \delta_{mk}$, i.e. first derivatives of the function $\varphi(y')$, whereas other coefficient depend also on second derivatives of $\varphi_0$ we have

$$K_1 \le C_4\lambda, \; K_2 \le C_5$$

and

$$|\underset{\sim}{k}_1(\hat{\underset{\sim}{w}}_\lambda, s_\lambda)|^{(\alpha,\alpha/2)}_{\mathbb{D}_T(\xi_0)} \le (C_4\lambda + \varepsilon)(\langle \hat{\underset{\sim}{w}}_\lambda \rangle^{(2+\alpha,1+\alpha/2)}_{\mathbb{D}_T}$$

$$+ \langle \nabla \hat{s}_\lambda \rangle^{(\alpha,\alpha/2)}_{\mathbb{D}_T}) + C_6(\varepsilon)\left( |(\hat{\underset{\sim}{w}}_\lambda)|_{\mathbb{D}_T} + |\nabla \hat{s}_\lambda|_{\mathbb{D}_T} \right).$$

Similar arguments are applicable to other operators $k_i$. In particular, $k_2(\hat{\underset{\sim}{w}}_\lambda)$ and the expression

$$\underset{\sim}{k}_6(\hat{\underset{\sim}{w}}_\lambda, \hat{s}_\lambda) = (I - \mathcal{J})(\nu\nabla^2 \hat{\underset{\sim}{w}}_\lambda - \nabla \hat{s}_\lambda) - \mathcal{J}\underset{\sim}{k}_1(\hat{\underset{\sim}{w}}_\lambda, \hat{s}_\lambda)$$

which is a part of the right-hand side in (2.13), are estimated precisely as $\underset{\sim}{k}_1$, namely,

$$|\nabla k_2(\hat{\underset{\sim}{w}}_\lambda)|^{(\alpha,\alpha/2)}_{\mathbb{D}_T} + |k_2(\hat{\underset{\sim}{w}}_\lambda)|^{\left(1+\alpha, \frac{1+\alpha}{2}\right)}_{\mathbb{R}_T} + |\underset{\sim}{k}_6(\hat{\underset{\sim}{w}}_\lambda, \hat{s}_\lambda)|^{(\alpha,\alpha/2)}_{\mathbb{D}_T}$$

$$\le (C_7\lambda + \varepsilon)\left( \langle \hat{\underset{\sim}{w}}_\lambda \rangle^{(2+\alpha,1+\alpha/2)}_{\mathbb{D}_T} + \langle \nabla \hat{s}_\lambda \rangle^{(\alpha,\alpha/2)}_{\mathbb{D}_T} \right. \tag{3.4}$$

$$\left. + C_8(\varepsilon)(|\hat{\underset{\sim}{w}}_\lambda|_{\mathbb{D}_T} + |\nabla \hat{s}_\lambda)|_{\mathbb{D}_T} \right).$$

Now, since $\underset{\sim}{n}_{00} = (0,0,-1)$ in the coordinate system $\{y_i\}$, it is seen from (2.3) that the components of $\hat{n}_0 - n_{00}$ and of

$$\Pi_{00}\hat{n}_0 = \hat{n}_0 - n_{00}(n_{00}\cdot\hat{n}_0) = \hat{n}_0 - n_{00} + n_{00}(\hat{n}_0 - n_{00}\cdot\hat{n}_0)$$

are linear combinations of functions, proportional to $\varphi_{y_1}$, or $\varphi_{y_2}$. Hence,

$$|k_3(\hat{w}_\lambda)|_{R_T}^{\left(1+\alpha,\frac{1+\alpha}{2}\right)} + |k_4(\hat{w}_\lambda)|_{R_T}^{\left(1+\alpha,\frac{1+\alpha}{2}\right)}$$

$$\le (C_9\lambda + \varepsilon)\langle\hat{w}_\lambda\rangle_{D_T}^{(2+\alpha,1+\alpha/2)} + C_{10}(\varepsilon)|\hat{w}_\lambda|_{D_T}.$$

In $k_5(\hat{w}_\lambda)$, one of the terms contains the difference $\hat{\sigma}(z) - \hat{\sigma}(0)$. Clearly,

$$|\hat{\sigma}(z) - \hat{\sigma}(0)| \le |\hat{\sigma}|_{R^2}^{(\alpha)}(2\lambda)^\alpha \qquad (|z| \le 2\lambda)$$

Another term is equal to

$$\hat{\sigma}(z)\left[(\hat{n}_0 - n_{00})\cdot\Delta\hat{w}_\lambda + n_{00}\cdot(\Delta - \Delta')\hat{w}_\lambda\right].$$

According to (2.6), leading coefficients of the operator $\Delta - \Delta' = \Delta''$ are equal to $-(1 + |\nabla\varphi|)^{-1}\varphi_{y_\alpha}\varphi_{y_\beta}$, so this term also can be estimated in the same way as $k_1$ or $\nabla k_2$. Altogether, we have

$$|k_5(\hat{w}_\lambda)|_{R_T}^{(\alpha,\alpha/2)} \le (C_9\lambda^\alpha + \varepsilon)\langle\hat{w}_\lambda\rangle_{D_T}^{(2+\alpha,1+\alpha/2)} + C_{10}(\varepsilon)|\hat{w}_\lambda|_{D_T}.$$

Consider finally the expression

$$k_{7m}(\hat{w}_\lambda,\hat{s}_\lambda) = (I - \mathcal{J})\left(\nu\frac{\partial\hat{w}_\lambda}{\partial z_m} - e_m\hat{s}_\lambda\right) - \mathcal{R}k_{1m}$$

in the right-hand side of (2.14). We estimate the norm $|k_{7m}|_{D_T}^{(1+\alpha,\gamma)}$ by inequality (2.29) taking into account that $\mathcal{J}$ is independent of t. This gives

$$|\underset{\sim}{k}_{7m}|_{\mathbb{D}_T}^{(1+\alpha,\gamma)} \le C_{11}\left\{\lambda\left(\left|\frac{\partial \hat{\underset{\sim}{w}}_\lambda}{\partial z_m}\right|_{\mathbb{D}_T}^{(1+\alpha,\gamma)} + |\hat{s}_\lambda|_{\mathbb{D}_T}^{(1+\alpha,\gamma)}\right)\right.$$

$$\left. + \lambda^{1-\gamma}\left(\left\langle\frac{\partial \hat{\underset{\sim}{w}}_\lambda}{\partial z_m}\right\rangle_{t,\mathbb{D}_T}^{\left(\frac{1+\alpha-\gamma}{2}\right)} + \langle\hat{s}_\lambda\rangle_{\mathbb{D}_T}^{(1+\alpha,\gamma)}\right)\right\} \tag{3.5}$$

$$\le (C_{12}\lambda + \varepsilon)\langle\hat{\underset{\sim}{w}}_\lambda\rangle_{\mathbb{D}_T}^{(2+\alpha,1+\alpha/2)} + C_{11}\lambda|\hat{s}_\lambda|_{\mathbb{D}_T}^{(1+\alpha,\gamma)}$$

$$+ C_{13}(\varepsilon)|\hat{\underset{\sim}{w}}_\lambda|_{\mathbb{D}_T} + C_{11}\langle\hat{s}_\lambda\rangle_{t,\mathbb{D}_T}^{\left(\frac{1+\alpha-\gamma}{2}\right)}\lambda^{1-\gamma}.$$

Now we are able to bound the norms of functions $\underset{\sim}{h}^{(2)}$ and $\underset{\sim m}{h}^{(2)}$ given by (2.13) and (2.14). In virtue of (3.4) and of lemma 2.4, we have

$$|\underset{\sim}{h}^{(2)}|_{\mathbb{D}_T}^{(\alpha,\alpha/2)} \le C_{14}\left[|\hat{\underset{\sim}{h}}^{(1)}|_{\mathbb{D}_T}^{(\alpha,\alpha/2)}\right.$$

$$+ (\lambda + \varepsilon)(\langle\hat{\underset{\sim}{w}}_\lambda\rangle_{\mathbb{D}_T}^{(2+\alpha,1+\alpha/2)} + \langle\nabla\hat{s}_\lambda\rangle_{\mathbb{D}_T}^{(\alpha,\alpha/2)} \tag{3.6}$$

$$\left. + C_8(\varepsilon)(|\hat{\underset{\sim}{w}}_\lambda|_{\mathbb{D}_T} + |\nabla\hat{s}_\lambda|_{\mathbb{D}_T} + \lambda^{1-\alpha}\left(|\hat{j}|_{\mathbb{D}_T} + \langle\hat{j}\rangle_{t,\mathbb{D}_T}^{(\alpha/2)}\right)\right].$$

Before evaluating the norm $|\underset{\sim m}{h}^{(2)}|_{\mathbb{D}_T}^{(1+\alpha,\gamma)}$ we rewrite the last term in (2.14). As

$$\hat{j} = (\hat{s} - \nu\hat{\rho})\hat{\nabla}^2\hat{\zeta}_\lambda - \hat{\nabla}\zeta_\lambda\cdot\sum_{k,m=1}^{3}\frac{\partial}{\partial y_m}J_{mk}\hat{h}_k,$$

it follows that

$$I \equiv \int_{\mathbb{D}}\frac{\hat{j}(y,t)}{|z-y|}\,dy = -\sum_{k,m=1}^{3}\frac{\partial}{\partial z_m}\int_{\mathbb{D}}\frac{\hat{\hat{\nabla}}\zeta_\lambda J_{mk}\cdot\hat{\underset{\sim}{h}}_k}{|z-y|}\,dy$$

$$-\int_{\mathbb{R}}\sum_{k,m=1}^{3}\hat{\hat{\nabla}}\zeta_\lambda\cdot n_{00m}\,J_{mk}\hat{\underset{\sim}{h}}_k\frac{dy'}{|z-y'|}$$

$$+\int_{\mathbb{D}}\left[(\hat{s} - \nu\hat{\rho})\hat{\nabla}^2\hat{\zeta}_\lambda + \sum_{k,m=1}^{3}\frac{\partial\hat{\hat{\nabla}}\zeta_\lambda}{\partial y_m}\cdot J_{mk}\hat{\underset{\sim}{h}}_k\right]\frac{dy}{|z-y|}.$$

Lemma 2.4 (and a similar lemma for a single layer potential

$$\int_{\mathbb{R}} f|z - y'|^{-1}dy')$$

implies

$$|I|_{D_T}^{(1+\alpha,\gamma)} \le C_{15}\lambda^{-\gamma}\left[\sum_{k=1}^{3} \langle h_k\rangle_{t,Q(2\lambda)}^{\left(\frac{1+\alpha-\gamma}{2}\right)} + \langle s - \nu\rho\rangle_{t,Q(2\lambda)}^{\left(\frac{1+\alpha-\gamma}{2}\right)}\right].$$

This inequality and the estimate (3.5) yield

$$|h_m^{(2)}|_{D_T}^{(1+\alpha,\gamma)} \le C_{16}\left(\sum_{k=1}^{3} |\hat{\zeta}_\lambda \hat{h}_k|_{D_T}^{(1+\alpha,\gamma)} + \lambda^{1-\gamma}\langle h^{(3)}\rangle_{t,D_T}^{\left(\frac{1+\alpha-\gamma}{2}\right)}\right)$$

$$+ |k_{7m}|_{D_T}^{(1+\alpha,\gamma)} + |I|_{D_T}^{(1+\alpha,\gamma)} \le C_{17}\left[\sum_{k=1}^{3} |\zeta_\lambda \hat{h}_k|_{D_T}^{(1+\alpha,\gamma)}\right.$$

$$+ \lambda^{-\gamma}\langle h_k\rangle_{t,D_T}^{\left(\frac{1+\alpha-\gamma}{2}\right)} + (\lambda + \varepsilon)\left(\langle \hat{w}_\lambda\rangle_{D_T}^{(2+\alpha,1+\alpha/2)} + \langle \nabla \hat{s}_\lambda\rangle_{D_T}^{(\alpha,\alpha/2)}\right) \tag{3.7}$$

$$+ C_8(\varepsilon)(|\hat{w}_\lambda|_{D_T} + |\nabla \hat{s}_\lambda|_{D_T} + \lambda^{1-\gamma}\left(\sum_{k=1}^{3} \langle \frac{\partial w_\lambda}{\partial z_k}\rangle_{t,D_T}^{\left(\frac{1+\alpha-\gamma}{2}\right)}\right.$$

$$\left.+ \lambda^{-1}\langle s\rangle_{t,Q(2\lambda)}^{\left(\frac{1+\alpha-\gamma}{2}\right)} + \langle h'\rangle_{t,Q(2\lambda)}^{\left(\frac{1+\alpha-\gamma}{2}\right)} + \lambda^{-1}\langle\rho\rangle_{t,Q(2\lambda)}^{\left(\frac{1+\alpha-\gamma}{2}\right)}\right].$$

We apply now theorem 3.1 to problem (2.9). According to (3.2),

$$\langle \hat{w}_\lambda\rangle_{D_T}^{(2+\alpha,1+\alpha/2)} + \langle \nabla \hat{s}_\lambda\rangle_{D_T}^{(\alpha,\alpha/2)} \le C_1(T)\left[\langle f^{(2)}\rangle_{D_T}^{(\alpha,\alpha/2)}\right.$$

$$+ \langle w_{0\lambda}\rangle_{D}^{(2+\alpha)} + \langle \nabla\rho^{(2)}\rangle_{D_T}^{(\alpha,\alpha/2)} + \langle\rho^{(2)}\rangle_{R_T}^{(1+\alpha,\frac{1+\alpha}{2})}$$

$$+ \langle h^{(2)}\rangle_{D_T}^{(\alpha,\alpha/2)} + \sum_{k=1}^{3} |\hat{h}_k^{(2)}|_{D_T}^{(1+\alpha,\gamma)} + \langle d^{(2)}\rangle_{R_T}^{(1+\alpha,\frac{1+\alpha}{2})} \tag{3.8}$$

$$\left.+ \langle b^{(2)}\rangle_{R_T}^{(1+\alpha,\frac{1+\alpha}{2})} + \langle B^{(2)}\rangle_{R_T}^{(\alpha,\alpha/2)}\right].$$

To evaluate the right-hand side, we make use of the above inequalities for $k_1$. In all these inequalities the sum

$$\langle \underset{\sim}{w}_\lambda \rangle_{D_T}^{(2+\alpha,1+\alpha/2)} + \langle \nabla s_\lambda \rangle_{D_T}^{(\alpha,\alpha/2)}$$

appears with the factor $\lambda + \varepsilon$ or $\lambda^\alpha + \varepsilon$. Both these numbers can be made small by an appropriate choice of $\lambda$ and $\varepsilon$. We take them so small that the cofficient by

$$\langle \underset{\sim}{w}_\lambda \rangle_{D_T}^{(2+\alpha,1+\alpha/2)} + \langle \nabla s \rangle_{D_T}^{(\alpha,\alpha/2)}$$

in the right-hand side of (3.8) gets less than 1/2, but we do not yet fix $\lambda$ definitely. This will give us the estimate

$$\begin{aligned} \langle \hat{\underset{\sim}{w}}_\lambda \rangle_{D_T}^{(2+\alpha,1+\alpha/2)} + \langle \hat{\hat{\nabla s}}_\lambda \rangle_{D_T}^{(\alpha,\alpha/2)} \le C_{18}\Big( & \langle \hat{\underset{\sim}{f}}^{(1)} \rangle_{D_T}^{(\alpha;\alpha/2)} \\ & + \langle \hat{\underset{\sim}{w}}_{0\lambda} \rangle_{D}^{(2+\alpha)} + \langle \nabla \hat{\rho}^{(1)} \rangle_{D_T}^{(\alpha,\alpha/2)} + \langle \hat{\rho}^{(1)} \rangle_{R_T}^{(1+\alpha,\frac{1+\alpha}{2})} \\ & \langle \hat{\underset{\sim}{d}}^{(1)} \rangle_{R_T}^{(1+\alpha,\frac{1+\alpha}{2})} + \langle \hat{b}^{(1)} \rangle_{R_T}^{(1+\alpha,\frac{1+\alpha}{2})} + \langle \hat{B}^{(1)} \rangle_{R_T}^{(\alpha,\alpha/2)} \\ & + |\hat{\underset{\sim}{h}}^{(1)}|_{D_T}^{(\alpha,\alpha/2)} + \sum_{m=1}^{3} |\zeta_\lambda \hat{\underset{\sim}{h}}_m|_{D_T}^{(1+\alpha,\gamma)} + Q_1 + Q_2 + \lambda |s_\lambda|_{D_T}^{(1+\alpha,\gamma)} \Big) \end{aligned} \tag{3.9}$$

where

$$Q_1 = \lambda^{1-\gamma} \sum_{k=1}^{3} \langle \frac{\partial \underset{\sim}{w}_\lambda}{\partial z_k} \rangle_{t,D_T}^{\left(\frac{1+\alpha-\gamma}{2}\right)} + \lambda^{-1} \langle s \rangle_{t,Q^{(2\lambda)}}^{\left(\frac{1+\alpha-\gamma}{2}\right)} + |\hat{w}_\lambda|_{D_T} + |\nabla \hat{s}_\lambda|_{D_T},$$

$$Q_2 = \lambda^{-\gamma} \Bigg( \langle \rho \rangle_{t,Q^{(2\lambda)}}^{\left(\frac{1+\alpha-\gamma}{2}\right)} + \sum_{k=1}^{3} \langle \underset{\sim}{h}_k \rangle_{t,Q^{(2\lambda)}}^{\left(\frac{1+\alpha-\gamma}{2}\right)} + \lambda \langle \underset{\sim}{h}' \rangle_{t,Q^{(2\lambda)}}^{\left(\frac{1+\alpha-\gamma}{2}\right)} \Bigg)$$

$$+ \left( |\hat{j}|_{D_T} + \langle \hat{j} \rangle_{t,D_T}^{(\alpha/2)} \right) \lambda^{1-\alpha}$$

As we can see from (2.7) and (2.12) the functions $\underset{\sim}{f}^{(1)}$, $\rho^{(1)}$, $\underset{\sim}{d}^{(1)}$,

$b^{(1)}$, $B^{(1)}$, $\underset{\sim}{h}^{(1)}$, $\underset{\sim}{h}'$ contain terms with $\underset{\sim}{w}$, $\frac{\partial \underset{\sim}{w}}{\partial x_1}$, $s_1$ multiplied by derivatives of $\zeta_\lambda$, and the sum of norms of these functions in (3.9) is less than

$$|\underset{\sim}{f}\zeta_\lambda|^{(\alpha,\alpha/2)}_{Q^{(2\lambda)}} + |\nabla\rho\zeta_\lambda|^{(\alpha,\alpha/2)}_{Q^{(2\lambda)}} + |\rho\zeta_\lambda|^{(1+\alpha,\frac{1+\alpha}{2})}_{S^{(2\lambda)}}$$

$$+ |\underset{\sim}{d}\zeta_\lambda|^{(1+\alpha,\frac{1+\alpha}{2})}_{S^{(2\lambda)}} + |\zeta_\lambda \underset{\sim}{h}|^{(\alpha,\alpha/2)}_{Q^{(2\lambda)}} + |b\zeta_\lambda|^{(1+\alpha,\frac{1+\alpha}{2})}_{S^{(2\lambda)}}$$

$$+ |B\zeta_\lambda|^{(\alpha,\alpha/2)}_{S^{(2\lambda)}} + C_{19}\lambda^{-\alpha-2}\left(|\underset{\sim}{w}|^{(\alpha,\alpha/2)}_{Q^{(2\lambda)}} + \sum_{m=1}^{3} |\underset{\sim}{w}_{\xi_m}|^{(\alpha,\alpha/2)}_{Q^{(2\lambda)}} + |s|^{(\alpha,\alpha/2)}_{Q^{(2\lambda)}}\right)$$

where $S^{(2\lambda)} = S_T \cap \partial Q^{(2\lambda)}$. The function j (2.12) also consists of such terms. Therefore, the inequality (3.9) implies

$$\langle \underset{\sim}{w}\rangle^{(2+\alpha,1+\alpha/2)}_{Q^{(2\lambda)}} + |\nabla s|^{(\alpha,\alpha/2)}_{Q^{(2\lambda)}} \le \langle \underset{\sim}{w}\rangle^{(2+\alpha,1+\alpha/2)}_{Q^{(2\lambda)}} + |\nabla s_\lambda|^{(\alpha,\alpha/2)}_{Q^{(2\lambda)}}$$

$$\le C_{20}(T)\left(|\underset{\sim}{f}\zeta_\lambda|^{(\alpha,\alpha/2)}_{Q^{(2\lambda)}} + |\nabla(\rho\zeta_\lambda)|^{(\alpha,\alpha/2)}_{Q^{(2\lambda)}} + |\rho\zeta_\lambda|^{(1+\alpha,\frac{1+\alpha}{2})}_{S^{(2\lambda)}}\right.$$

$$+ |\underset{\sim}{d}\zeta_\lambda|^{(1+\alpha,\frac{1+\alpha}{2})}_{S^{(2\lambda)}} + |b\zeta_\lambda)|^{(1+\alpha,\frac{1+\alpha}{2})}_{S^{(2\lambda)}} + |B\zeta_\lambda|^{(\alpha,\alpha/2)}_{S^{(2\lambda)}} \tag{3.10}$$

$$\left. + |\underset{\sim}{h}\zeta_\lambda|^{(\alpha,\alpha/2)}_{Q^{(2\lambda)}} + \sum_{k=1}^{3} |\zeta_\lambda \underset{\sim}{h}_k|^{(1+\alpha,\gamma)}_{Q^{(2\lambda)}} + \lambda|s\zeta_\lambda|^{(1+\alpha,\gamma)}_{Q^{(2\lambda)}} + Q_3\right)$$

with

$$Q_3 = \lambda^{-\alpha-2}\left(|\underset{\sim}{w}|^{(\alpha,\alpha/2)}_{Q^{(2\lambda)}} + \sum_{m=1}^{3} |\underset{\sim}{w}_{\xi_m}|^{(\alpha,\alpha/2)}_{Q^{(2\lambda)}} + |s|^{(\alpha,\alpha/2)}_{Q^{(2\lambda)}}\right.$$

$$\left. + |\nabla s|^{\left(\frac{1+\alpha-\gamma}{2}\right)}_{Q^{(2\lambda)}} + \langle \underset{\sim}{w}_{\xi_m}\rangle^{\left(\frac{1+\alpha-\gamma}{2}\right)}_{t,Q^{(2\lambda)}} + \langle s\rangle^{\left(\frac{1+\alpha-\gamma}{2}\right)}_{t,Q^{(2\lambda)}}\right).$$

Similar estimate (but without boundary norms) can be obtained in the case $\mathrm{dist}(x_0,\Gamma)>\lambda/2$, when we can consider $\underset{\sim}{w}\zeta_{\lambda/4}$, $s\zeta_{\lambda/4}$ as a solution of the Cauchy problem (2.15). Returning to inequality (3.3) we have started with we see that in virtue of (3.10)

$$\langle \underset{\sim}{w}\rangle^{(2+\alpha,1+\alpha/2)}_{Q_T} + |\nabla s|^{(\alpha,\alpha/2)}_{Q_T} \le C_{21}\Bigg[\mathcal{F}(T) + \lambda|s|^{(1+\alpha,\gamma)}_{Q_T}$$

$$+ \lambda^{-\alpha-2}\Bigg(\sum_{|k|\le 2} |D^k_\xi \underset{\sim}{w}|_{Q_T} + |\underset{\sim}{w}_t|_{Q_T} + \sup_{t\le T} |\underset{\sim}{w}|^{(1+\alpha)}_{\Omega} \qquad (3.11)$$

$$+ \sum_{m=1}^{3} \langle \underset{\sim}{w}_{\xi_m}\rangle^{\left(\frac{1+\alpha-\gamma}{2}\right)}_{t,Q_T} + |\nabla s|_{Q_T} + |s|_{Q_T} + \langle s\rangle^{\left(\frac{1+\alpha-\gamma}{2}\right)}_{t,Q_T}\Bigg)\Bigg]$$

where $\mathcal{F}(T)$ is the sum of norms in the right-hand side of (1.10). To estimate s we consider it as a solution of the Dirichlet problem

$$\nabla^2 s = \nabla\cdot(\underset{\sim}{f} - \underset{\sim}{w}_t + \nu\nabla^2\underset{\sim}{w}) = \nabla\cdot(\nu\nabla\rho - \underset{\sim}{h}) \equiv \nabla\cdot\sum_{m=1}^{3}\frac{\partial \underset{\sim}{H}_m}{\partial\xi_m} \quad (\xi\in\Omega),$$
$$(3.12)$$
$$s = -b + 2\nu\underset{\sim}{n}_0\cdot\frac{\partial\underset{\sim}{w}}{\partial n} - \int_0^t(\sigma\underset{\sim}{n}_0\cdot\Delta\underset{\sim}{w} + B)d\tau \equiv s_0 \quad (\xi\in\Gamma)$$

with $\underset{\sim}{H}_m = -\nu\underset{\sim}{e}_m\rho - \underset{\sim}{h}_m$. It was proved in [S6] that

$$|s(\cdot,t)|_\Omega \le C_{22}\sum_{m=1}^{3}\langle \underset{\sim}{H}_m\rangle^{(\gamma)}_\Omega + |s_0(\cdot,t)|_\Gamma,$$

$$|s(\cdot,t)|^{(\gamma)}_\Omega \le C_{23}\sum_{m=1}^{3}\langle \underset{\sim}{H}_m\rangle^{(\gamma)}_\Omega + |s_0(\cdot,t)|^{(\gamma)}_\Gamma,$$

The first inequality, applied to $s(\xi,t)$ and to $s(\xi,t) - s(\xi,t-\tau)$, gives

$$|s|_{Q_T} + \langle s\rangle_{t,Q_T}^{\left(\frac{1+\alpha-\gamma}{2}\right)} \le C_{24}\left(|\rho|_{Q_T}^{(1+\alpha,\gamma)} + \sum_{m=1}^{3} |\underset{\sim}{h}_m|_{Q_T}^{(1+\alpha,\gamma)}\right.$$

$$+ |b|_{S_T} + \langle b\rangle_{t,S_T}^{\left(\frac{1+\alpha-\gamma}{2}\right)} + \sum_{m=1}^{3}\left(\left|\frac{\partial \underset{\sim}{w}}{\partial \xi_m}\right|_{Q_T} + \left\langle\frac{\partial \underset{\sim}{w}}{\partial \xi_m}\right\rangle_{t,Q_T}^{\left(\frac{1+\alpha-\gamma}{2}\right)}\right) \tag{3.13}$$

$$\left.+ T^{\frac{1-\alpha+\gamma}{2}}\left(|B|_{S_T} + \sum_{|k|=2} |D^k \underset{\sim}{w}|_{Q_T}\right)\right).$$

If we apply the second inequality to the difference $s(\xi,t) - s(\xi,t-\tau)$ we easily obtain

$$|s|_{Q_T}^{(1+\alpha,\gamma)} \le C_{25}\left(|\rho|_{Q_T}^{(1+\alpha,\gamma)} + \sum_{m=1}^{3} |\underset{\sim}{h}_m|_{Q_T}^{(1+\alpha,\gamma)}\right.$$

$$\left.+ |b|_{S_T}^{(1+\alpha,\gamma)} + \sum_{m=1}^{3}\left|\frac{\partial \underset{\sim}{w}}{\partial \xi_m}\right|_{Q_T}^{(1+\alpha,\gamma)} + |J|_{S_T}^{(1+\alpha,\gamma)}\right)$$

where

$$J(\xi,t) = \int_0^t (\sigma \underset{\sim}{n}_0 \cdot \Delta \underset{\sim}{w} + B)d\tau.$$

In virtue of lemma 2.2,

$$|J|_{S_T}^{(1+\alpha,\gamma)} \le C_{26}\left(\langle \nabla_\tau J\rangle_{t,S_T}^{(\alpha/2)} + \langle J\rangle_{t,S_T}^{\left(\frac{1+\alpha}{2}\right)}\right) \le C_{26}(\langle \nabla_\tau J\rangle_{t,S_T}^{(\alpha/2)} + |J_t|_{S_T})$$

($\nabla_\tau$ is the gradient on $\Gamma$). From the boundary condition (3.12) we find

$$\langle \nabla_\tau J\rangle_{t,S_T}^{(\alpha/2)} \le \langle \nabla_\tau b\rangle_{t,S_T}^{(\alpha/2)} + \langle \nabla s\rangle_{t,Q_T}^{(\alpha/2)} + 2\nu\left\langle \nabla_\tau\left(\frac{\partial \underset{\sim}{w}}{\partial n}\cdot \underset{\sim}{n}_0\right)\right\rangle_{t,S_T}^{(\alpha/2)}$$

and, as a conseqence,

$$|s|_{Q_T}^{(1+\alpha,\gamma)} \le C_{27}\left(|\nabla s|_{Q_T}^{(\alpha,\alpha/2)} + |\rho|_{Q_T}^{(1+\alpha,\gamma)} + \sum_{m=1}^{3} |\underset{\sim}{h}_m|_{Q_T}^{(1+\alpha,\gamma)}\right. \tag{3.14}$$

$$\left.+ |\underset{\sim}{w}|_{Q_T}^{(2+\alpha,1+\alpha/2)} + |b|_{S_T}^{\left(1+\alpha,\frac{1+\alpha}{2}\right)} + |B|_{S_T}^{(\alpha,\alpha/2)}\right)$$

Now we estimate norms of s in (3.11) by inequalities (3.13), (3.14), and

$$|\nabla s|_{Q_T} \le \varepsilon_1^{\alpha} \sup_{t\le T} \langle \nabla s\rangle_{\Omega}^{(\alpha)} + C_{28}\varepsilon_1^{-1}|s|_{Q_T}$$

and we take first $\lambda$, then $\varepsilon_1$ so small that

$$C_{21}C_{27}\lambda + + C_{21}\lambda^{-2-\alpha}\varepsilon_1^{\alpha} \le 1/2$$

(this is the second and the last restriction for $\lambda$ which is now fixed definitely). As a result, we obtain

$$\langle \underset{\sim}{w}\rangle_{Q_T}^{(2+\alpha,1+\alpha/2)} + |\nabla s|_{Q_T}^{(\alpha,\alpha/2)} \le C_{29}(\mathcal{F}(T) + Q[\underset{\sim}{w}]) \tag{3.15}$$

where

$$Q[\underset{\sim}{w}] = \sum_{m=1}^{3} \langle \underset{\sim}{w}_{\xi_m}\rangle_{t,Q_T}^{\left(\frac{1+\alpha-\gamma}{2}\right)} + |\underset{\sim}{w}_t|_{Q_T} + \sum_{|k|\le 2} |D_{\xi}^{k}\underset{\sim}{w}|_{Q_T}$$

can be estimated by interpolation inequality $(2.16_2)$ as follows:

$$\langle \underset{\sim}{w}_{\xi_m}\rangle_{t,Q_T}^{\left(\frac{1+\alpha-\gamma}{2}\right)} \le \left(\langle \underset{\sim}{w}_{\xi_m}\rangle_{t,Q_T}^{\left(\frac{1+\alpha}{2}\right)}\right)^{\frac{1+\alpha-\gamma}{1+\alpha}} \cdot \left(2|\underset{\sim}{w}_{\xi_m}|_{Q_T}\right)^{\frac{\gamma}{1+\alpha}},$$

$$\sum_{|k|\le 2} |D_{\xi}^{k}\underset{\sim}{w}|_{Q_T} \le \varepsilon_2^{\alpha}\sup_{t\le T} \langle \underset{\sim}{w}\rangle_{\Omega}^{(2+\alpha)} + C_{30}(\varepsilon_2)|\underset{\sim}{w}|_{Q_T}, \tag{3.16}$$

$$|\underset{\sim}{w}_t|_{Q_T} \le \varepsilon_2^{\alpha}\langle \underset{\sim}{w}\rangle_{t,Q_T}^{(\alpha/2)} + C_{31}(\varepsilon_2)(|\underset{\sim}{w}|_{Q_T} + |\underset{\sim}{w}_t(\cdot,0)|_{\Omega}).$$

Altogether, there holds

$$Q[\underset{\sim}{w}] \le 3\varepsilon_2^{\alpha}\langle \underset{\sim}{w}\rangle_{Q_T}^{(2+\alpha,1+\alpha/2)} + C_{32}(\varepsilon_2)(|\underset{\sim}{w}|_{Q_T} + |\underset{\sim}{w}_t(\cdot,0)|_{\Omega}) \tag{3.17}$$

We write the function $\underset{\sim}{w}_t(\xi,0)$ as

$$\underset{\sim}{w}_t(\xi,0) = \nu\nabla^2\underset{\sim}{w}_0 + \underset{\sim}{f}(\xi,0) - \nabla s(\xi,0)$$

and observe that $s(\xi,0) = s_0(\xi)$ is a solution of the Dirichlet problem (3.12) for $t = 0$, i.e.

$$\nabla^2 s_0 = \nabla\cdot(\nu\nabla\rho(\xi,0) - \underset{\sim}{h}(\xi,0)),\ s_0|_{\Gamma} = -b(\xi,0) + 2\nu\underset{\sim}{n}_0\cdot\frac{\partial \underset{\sim}{w}_0}{\partial n}.$$

It follows from general results of S.Agmon, A. Douglis and L. Nirenberg (see [ADN]) that

$$|\nabla s_0|_{\Omega}^{(\alpha)} = C_{33}\left(|\nu\nabla\rho(\cdot,0) - h(\cdot,0)|_{\Omega}^{(\alpha)} + |b(\cdot,0)|_{\Gamma}^{(1+\alpha)}\right.$$

$$\left. + 2\nu\left|n_0\cdot\frac{\partial w_0}{\partial n}\right|_{\Gamma}^{(1+\alpha)}\right) \le C_{34}\mathcal{F}(T)$$

so (3.15) and (3.17) imply

$$\langle \underset{\sim}{w}\rangle_{Q_T}^{(2+\alpha,1+\alpha/2)} + \langle\nabla s\rangle_{Q_T}^{(\alpha,\alpha/2)} \le C_{35}(\mathcal{F}(T) + |\underset{\sim}{w}|_{Q_T}). \tag{3.18}$$

It remains to obtain an estimate for $|\underset{\sim}{w}|_{Q_T}$. We observe that in all our calculations the constants $C_i$ were non-decreasing functions of T and the inequalities (3.18) and (3.16) hold also in any smaller cylinder $Q_t$, $t \le T$, with the same $C_{31}$, and $C_{35}$. It follows from these inequalities that

$$|\underset{\sim}{w}|_{Q_t} \le |\underset{\sim}{w}_0|_{\Omega} + \int_0^t |\underset{\sim}{w}_t(\cdot,\tau)|_{\Omega}d\tau \le C_{36}\left(\mathcal{F}(t) + \int_0^t |\underset{\sim}{w}|_{Q_T}d\tau\right),$$

and by the Gronwall lemma

$$|\underset{\sim}{w}|_{Q_T} \le C_{37}(t)\mathcal{F}(t).$$

This inequality together with (3.13)-(3.18) evidently imply a priori estimate (1.10) that is finally established.

## §4. Proof of the solvability of problem (1.8).

In this section we establish the solvability of the problem (1.8) and complete the proof of theorem 1.1. Our proof shall be based on theorem 1.1′. This makes it possible to reduce a general problem to the particular case $\underset{\sim}{f} = 0$, $\rho = 0$, $\underset{\sim}{w}_0 = 0$, $\underset{\sim}{d} = 0$, $b = 0$ by introducing new unknown functions $\underset{\sim}{v} = \underset{\sim}{w} - \underset{\sim}{u}$, $p = s - q$ where $(\underset{\sim}{u},q)$ is a solution of the same problem (1.8) with $\sigma = 0$. Moreover, it can be assumed without restriction of generality that $B|_{t=0} = 0$, since otherwise we could use the identity

$$\int_0^t B d\tau = \int_0^t \left(B - \frac{\partial A}{\partial \tau}\right) d\tau + A(\xi,t)$$

where $A \in C^{2+\alpha,1+\alpha/2}(S_T)$ satisfies the initial conditions

$$A\big|_{t=0} = 0, \quad \frac{\partial A}{\partial t}\bigg|_{t=0} = B\bigg|_{t=0}$$

and the inequality

$$|A|_{S_T}^{(2+\alpha,1+\alpha/2)} \le C_1 |B|_{S_T}^{(\alpha,\alpha/2)},$$

so that

$$B - \frac{\partial A}{\partial t} \in C^{(\alpha,\alpha/2)}(S_T), \qquad B - \frac{\partial A}{\partial t}\bigg|_{t=0} = 0.$$

A new term A in the boundary condition can be eliminated by the solution of problem (1.8) with $\sigma = 0$ and with

$$b = A,\ \underset{\sim}{f} = 0,\ \rho = 0,\ \underset{\sim}{w}_0 = 0,\ \underset{\sim}{d} = 0.$$

Thus, we consider the initial-boundary value problem

$$\begin{gathered} \underset{\sim}{v}_t - \nu\nabla^2 \underset{\sim}{v} + \nabla p = 0, \quad \nabla\cdot\underset{\sim}{v} = 0 \\ \underset{\sim}{v}\big|_{t=0} = 0, \\ \nu\Pi_0 S(\underset{\sim}{v})\underset{\sim}{n}_0 = 0, \\ \underset{\sim}{n}_0 \cdot T(\underset{\sim}{v},p)\underset{\sim}{n}_0 - \sigma(\xi)\underset{\sim}{n}_0\cdot\Delta\int_0^t \underset{\sim}{v} d\tau = \int_0^t B d\tau \end{gathered} \tag{4.1}$$

with

$$B \in C^{\alpha,\alpha/2}(S_T), \qquad B\big|_{t=0} = 0.$$

Denote by

$$\overset{\circ}{C}{}^{1,1/2}(\mathcal{G}_T), \quad \mathcal{G}_T = G \times (0,T), \quad G \in \mathbb{R}^n,$$

the space of all the elements of $C^{1,1/2}(\mathcal{G}_T)$ such that

$$\frac{\partial^a u}{\partial t^a}\bigg|_{t=0} = 0, \quad a = 0,\dots,\left[\frac{1}{2}\right].$$

and let $\tilde{C}^{1+\alpha}(Q_T)$ be the space of functions $q(x,t)$, $(x,t) \in Q_T$, with a finite norm

$$|q|_{\tilde{C}^{1+\alpha}(Q_T)} = |\nabla q|_{Q_T}^{(\alpha,\alpha/2)} + |q|_{Q_T}^{(\alpha,\alpha/2)} + |q|_{S_T}^{1+\alpha,\frac{1+\alpha}{2}}.$$

We construct a linear continuous operator

$$R:\ \overset{\circ}{C}{}^{\alpha,\alpha/2}(S_T) \to \overset{\circ}{C}{}^{2+\alpha,1+\alpha/2}(Q_T) \times \tilde{C}^{1+\alpha}(Q_T)$$

with the following properties:

a) $(\underset{\sim}{v},p) = RB$ is a solution of problem (4.1) with $B' = B + MB$ in place of B.

b) M is a linear continuous operator and the operator $I + M$ is invertible.

It follows that $(\underset{\sim}{v},p) = R(I + M)^{-1}B$ is a solution of (4.1). The uniqueness of the solution follows from a-priori estimate (1.10).

Let $\{\zeta_j\}_{j=1,\dots,N_\lambda}$ be a smooth partition of unity on $\Gamma$ subordinate to the covering of $\Gamma$ by the balls

$$K_{j,\lambda} = \{|\xi - \xi_j| < \lambda\},\ \xi_j \in \Gamma,\ \lambda < d/2,$$

and let $\eta_j$ be smooth functions with $\operatorname{supp}\eta_j \subset K_{j,2\lambda}$ such that $\eta_j\zeta_j = \zeta_j$. We assume that

$$|D^k\zeta_j(\xi)| + |D^k\eta_j(\xi)| \le c_2(|k|)\lambda^{-|k|}.$$

By $\varphi_j(y')$ we denote functions defining $\Gamma$ in neighbourhoods of $\xi_j$ by equations (2.1) in local cartesian coordinates $(y_1,y_2,y_3)$ at the point $\xi_j$, and we set $Z_j \equiv Z_{\xi_j}$, $D_j = D(\xi_j)$, $R_j = R(\xi_j) = \partial D_j$.

Consider the functions

$$\underset{\sim}{v}'(\xi,t) = \sum_{j=1}^{N_\lambda} \eta_j(\xi)\underset{\sim}{v}_j(\xi,t), \qquad p'(x,t) = \sum_{j=1}^{N_\lambda} \eta_j(\xi)p_j(\xi,t)$$

where $\underset{\sim}{v}_j(\xi,t) = \underset{\sim}{w}_j(Z_j(\xi),t)$, $p_j(\xi,t) = s_j(Z_j(\xi),t)$, and $(\underset{\sim}{w}_j(z,t),\ s_j(z,t))$ is a solution of the half-space problem

$$\begin{aligned}
&\underset{\sim}{w}_{jt} - \nu\nabla^2\underset{\sim}{w}_j + \nabla s_j = 0, \quad \nabla\cdot\underset{\sim}{w}_j = 0 \quad (z \in D_j),\\
&\underset{\sim}{w}_j\big|_{t=0} = 0,\\
&\nu\Pi_j S(\underset{\sim}{w}_j)\underset{\sim}{n}_0(\xi_j) = 0,\\
&\underset{\sim}{n}_0(\xi_j)\cdot T(\underset{\sim}{w}_j,s_j)\underset{\sim}{n}_0(\xi_j) - \sigma(\xi_j)\left(\underset{\sim}{n}_0(\xi_j)\cdot\Delta_j\int_0^t \underset{\sim}{w}_j\,d\tau\right)\\
&\quad = \int_0^t B_j\,d\tau \quad (z \in R_j)
\end{aligned} \tag{4.2}$$

where $\Pi_j$ is a projection onto $\mathbb{R}_j$, $\Delta_j$ is a Laplacian on $\mathbb{R}_j$ and

$$B_j(z,t) = B(\xi,t)\zeta_j(x,t)\Big|_{\xi=Z_j^{-1}(z)} \text{ for } |z| \le \lambda,$$

$$B_j(z,t) = 0 \text{ for } |z| \ge \lambda,\ z \in \mathbb{R}_j.$$

In the coordinates $\xi = Z_j^{-1}(z)$ the equations (4.2) take the form

$$\underset{\sim}{v}_{jt} - \nu\nabla_j^2 \underset{\sim}{v}_j + \nabla_j p_j = 0, \quad \nabla_j \cdot \underset{\sim}{v}_j = 0,$$

$$\underset{\sim}{v}_j\Big|_{t=0} = 0,$$

$$\Pi_j S^{(j)}(\underset{\sim}{v}_j)\underset{\sim}{n}_0(\xi_j) = 0,$$

$$\underset{\sim}{n}_0(\xi_j)\cdot T^{(j)}(\underset{\sim}{v}_j,p_j)\underset{\sim}{n}_0(\xi_j) - \sigma(\xi_j)\left(\underset{\sim}{n}_0(\xi_j)\cdot\Delta_j\int_0^t \underset{\sim}{v}_j d\tau\right) = \int_0^t B\zeta_j d\tau$$

with

$$\nabla_j = \left(\mathcal{J}_j^{-1}\right)^*\nabla, \quad T^{(j)}(\underset{\sim}{v},p) = -pI + \nu S^{(j)}(\underset{\sim}{v}),$$

$$S_{lk}^{(j)} = \sum_{m=1}^{3}\left(J_j^{mk}\frac{\partial v_l}{\partial \xi_m} + J_j^{ml}\frac{\partial v_k}{\partial \xi_m}\right),$$

$\mathcal{J}_j^{-1}$ is the Jacobian matrix of the transformation $Z_j^{-1}$ with elements $J_j^{mk}$. Next, we define $(\underset{\sim}{v}'',p'')$ as a solution of the problem

$$\begin{gathered}\underset{\sim}{v}''_t - \nu\nabla^2\underset{\sim}{v}'' + \nabla p'' = \underset{\sim}{f}, \quad \nabla\cdot\underset{\sim}{v}'' = \rho, \quad \underset{\sim}{v}''\Big|_{t=0} = 0. \\ T(\underset{\sim}{v}'',p'')\underset{\sim}{n}_0\Big|_{\xi\in\Gamma} = \underset{\sim}{d}\end{gathered} \tag{4.3}$$

where

$$\underset{\sim}{f} = \sum_{j=1}^{N_\lambda}\eta_j\left[\nu(\nabla^2 - \nabla_j^2)\underset{\sim}{v}_j + (\nabla_j - \nabla)p_j\right] + \sum_{j=1}^{N_\lambda}\left[2\nu(\nabla\eta_j\cdot\nabla)\underset{\sim}{v}_j\right.$$

$$\left. + \nu\underset{\sim}{v}_j\nabla^2\eta_j - p_j\nabla\eta_j\right],$$

$$\rho = -\sum_{j=1}^{N_\lambda}\nabla\cdot(\eta_j\underset{\sim}{v}_j) = -\sum_j\nabla\eta_j\cdot\underset{\sim}{v}_j - \sum_j\eta_j(\nabla - \nabla_j)\cdot\underset{\sim}{v}_j,$$

$$\underset{\sim}{d} = -\nu\sum_{j=1}^{N_\lambda}\left[(\underset{\sim}{v}_j\cdot\underset{\sim}{n}_0)\nabla\eta_j + \frac{\partial\eta_j}{\partial n}\underset{\sim}{v}_j\right] + \sum_j\eta_j\left[T^{(j)}(\underset{\sim}{v}_j,p_j)\underset{\sim}{n}_0(\xi_j)\right.$$

$$-T(\underset{\sim}{v}_j,p_j)\underset{\sim}{n}_0\Big] + (\underset{\sim}{n}_0(\xi) - n_0(\xi_j))(\underset{\sim}{n}_0(\xi_j)\cdot T^{(j)}(\underset{\sim}{v}_j,p_j)\underset{\sim}{n}_0(\xi_j))$$

and we set

$$RB = (\underset{\sim}{v},p) = (\underset{\sim}{v}' + \underset{\sim}{v}'', p' + p'').$$

It is not hard to verify that $(\underset{\sim}{v},p)$ satisfy a homogeneous Stokes system, a homogeneous initial condition, and the boundary condition

$$T(\underset{\sim}{v},p)\underset{\sim}{n}_0(\xi) - \sigma(\xi)\underset{\sim}{n}_0(\xi)\left(\underset{\sim}{n}_0\cdot\Delta\int_0^t \underset{\sim}{v}d\tau\right) = \underset{\sim}{n}_0(\xi)\int_0^t Bd\tau$$

$$- \sigma(\xi)\underset{\sim}{n}_0(\xi)\left\{\left(\underset{\sim}{n}_0\cdot \Delta\int_0^t \underset{\sim}{v}''d\tau\right) + \underset{\sim}{n}_0(\xi)\cdot\sum_{j=1}^{N_\lambda}\int_0^t(\Delta\eta_j\underset{\sim}{v}_j - \eta_j\Delta\underset{\sim}{v}_j)d\tau\right.$$

$$\left. + \sum_{j=1}^{N_\lambda}\eta_j(\xi)\int_0^t\left[\underset{\sim}{n}_0(\xi)\cdot\Delta\underset{\sim}{v}_j - \underset{\sim}{n}_0(\xi_j)\cdot\Delta_j\underset{\sim}{v}_j\right]d\tau\right\}$$

$$- \underset{\sim}{n}_0(\xi)\sum_{j=1}^{N_j}\eta_j(\xi)[\sigma(\xi) - \sigma(\xi_j)]\underset{\sim}{n}_0(\xi_j)\cdot\Delta_j\int_0^t\underset{\sim}{v}_j d\tau.$$

This means that

$$MB = -\sigma\underset{\sim}{n}_0\cdot\Delta\underset{\sim}{v}'' - \sigma\underset{\sim}{n}_0\cdot\sum_{j=1}^{N_\lambda}[\Delta(\eta_j\underset{\sim}{v}_j) - \eta_j\Delta\underset{\sim}{v}_j]d\tau$$

$$- \sigma\sum_{j=1}^{N_\lambda}\eta_j[\underset{\sim}{n}_0(\xi)\cdot\Delta\underset{\sim}{v}_j - \underset{\sim}{n}_0(\xi_j)\cdot\Delta_j\underset{\sim}{v}_j] \tag{4.4}$$

$$- \sum_{j=1}^{N_\lambda}\eta_j[\sigma(\xi) - \sigma(\xi_j)]\underset{\sim}{n}_0(\xi_j)\cdot\Delta_j\underset{\sim}{v}_j.$$

The objective of the following arguments is the estimate

$$|MB|_{S_t}^{(\alpha,\alpha/2)} \le (C_3\lambda^\alpha + \varepsilon)|B|_{S_t}^{(\alpha,\alpha/2)} + C_4(\lambda,\varepsilon)\int_0^t|B|_{S_\tau}^{(\alpha,\alpha/2)}d\tau. \tag{4.5}$$

for arbitrary $t \le T$ and $\varepsilon \in (0,1)$, with a constant $C_3$ independent of $\lambda$ and $\varepsilon$. If $C_3\lambda^\alpha + \varepsilon < 1$, this estimate implies the existence of a bounded $(I + M)^{-1}$.

We turn to the half-space problem (4.2). Since $\underset{\sim}{w}_j(z,0) = \underset{\sim}{w}_{jt}(z,0) = 0$,

$s_j(z,0) = 0$, it follows from (3.2) that

$$|\underset{\sim}{w}_j|^{(2+\alpha,1+\alpha/2)}_{\mathbb{D}^{(j)}_T} + |\nabla s_j|^{(\alpha,\alpha/2)}_{\mathbb{D}^{(j)}_T} \le C_5(T)|B_j|^{(\alpha,\alpha/2)}_{\mathbb{R}^{(j)}_T} \tag{4.6}$$

where $\mathbb{D}^{(j)}_T = \mathbb{D}_j \times (0,T)$, $\mathbb{R}^{(j)}_T = R_j \times (0,T)$ and $C_5(T)$ is a non-decreasing function of T. Moreover, if we consider $s_j$ as a solution of the Dirichlet problem

$$\nabla^2 s_j = 0 \qquad (z \in \mathbb{D}_j),$$

$$s_j = 2\nu \underset{\sim}{n}_0(\xi_j)\cdot \frac{\partial \underset{\sim}{w}_j}{\partial n} - \sigma(\xi_j)\int_0^t (B_j - \underset{\sim}{n}_0(\xi_j)\cdot\Delta_j \underset{\sim}{w}_j)d\tau, \qquad (z \in R_j)$$

we easily obtain the inequality

$$|s_j|^{(\alpha,\alpha/2)}_{\mathbb{D}^{(j)}_t} \le C_6\Big[2\nu\Big|\frac{\partial \underset{\sim}{w}_j}{\partial n}\Big|^{(\alpha,\alpha/2)}_{\mathbb{R}^{(j)}_t} + \sigma(\xi_j)\int_0^t |B_j - \underset{\sim}{n}_0(\xi_j)\cdot\Delta_j \underset{\sim}{w}_j|^{(\alpha,\alpha/2)}_{\mathbb{R}^{(j)}_\tau} d\tau + \sigma(\xi_j)t^{1-\alpha/2}|B_j - \underset{\sim}{n}_0(\xi)\cdot\Delta_j \underset{\sim}{w}_j|_{\mathbb{R}^{(j)}_t}\Big]. \tag{4.7}$$

In virtue of interpolation inequalities (2.20) (with the constant independent of t since all the functions admit zero extension into the half-space $t \le 0$ with the preservation of class) we have

$$\Big|\frac{\partial \underset{\sim}{w}_j}{\partial n}\Big|^{(\alpha,\alpha/2)}_{\mathbb{R}^{(j)}_t} + |\Delta_j \underset{\sim}{w}_j|_{\mathbb{R}^{(j)}_t} \le \varepsilon_1 |\underset{\sim}{w}_j|^{(2+\alpha,1+\alpha/2)}_{\mathbb{D}^{(j)}_t} + C_7(\varepsilon_1)|\underset{\sim}{w}_j|_{\mathbb{D}^{(j)}_t} \le \varepsilon_1|\underset{\sim}{w}_j|^{(2+\alpha,1+\alpha/2)}_{\mathbb{D}^{(j)}_t} + C_7(\varepsilon_1)\int_0^t |\underset{\sim}{w}_j|^{(2+\alpha,1+\alpha/2)}_{\mathbb{D}^{(j)}_\tau} d\tau, \quad \forall\varepsilon_1 \in (0,1). \tag{4.8}$$

We also estimate the function $B_j$ by the interpolation inequality

$$|B_j|_{\mathbb{R}^{(j)}_t} \le \varepsilon_2|B_j|^{(\alpha,\alpha/2)}_{\mathbb{R}^{(j)}_t} + C_8(\varepsilon_2)\int_{\mathbb{R}^{(j)}_t} |B_j| dz'\,dt \le \varepsilon_2|B_j|^{(\alpha,\alpha/2)}_{\mathbb{R}^{(j)}_t} + C_8(\varepsilon_2)\int_0^t |B_j|^{(\alpha,\alpha/2)}_{\mathbb{R}^{(j)}_\tau} d\tau, \quad \forall\varepsilon_2 \in (0,1). \tag{4.9}$$

It follows from (4.7) – (4.9) that

$$|s_j|^{(\alpha,\alpha/2)}_{D_t^{(j)}} \le \varepsilon_3 |B_j|^{(\alpha,\alpha/2)}_{R_t^{(j)}} + C_9(\varepsilon_3)\int_0^t |B_j|^{(\alpha,\alpha/2)}_{R_\tau^{(j)}} d\tau \tag{4.10}$$

with a properly chosen $\varepsilon_3$ that can be made as small as we wish by taking $\varepsilon_1$ and $\varepsilon_2$ small. In the same way it can also be shown that

$$\langle s_j\rangle^{(\beta)}_{t,D_t^{(j)}} \le \varepsilon_4 |B_j|^{(\alpha,\alpha/2)}_{R_t^{(j)}} + C_{10}(\varepsilon_4)\int_0^t |B_j|^{(\alpha,\alpha/2)}_{R_\tau^{(j)}} d\tau \tag{4.11}$$

where $\beta < \frac{1+\alpha}{2}$ and $\varepsilon_4 \in (0,1)$. Finally, applying the maximum principle to the difference $s_j(z,t) - s_j(z,t-\tau)$ we easily obtain

$$\langle s_j\rangle^{(\frac{1+\alpha}{2})}_{t,D_t^{(j)}} \le 2\nu\langle\frac{\partial \tilde{w}_j}{\partial n}\rangle^{(\frac{1+\alpha}{2})}_{t,R_t^{(j)}} + \sigma(\xi_j)t^{\frac{1-\alpha}{2}}|B - \tilde{n}_0(\xi)\cdot\Delta\tilde{w}_j|_{R_T^{(j)}}$$

which yields, together with (4.6), an estimate of $C^{1+\alpha,\frac{1+\alpha}{2}}$ - norm of $s_j$:

$$|s_j|^{(1+\alpha,\frac{1+\alpha}{2})}_{D_t^{(j)}} \le C_{11}(T)|B|^{(\alpha,\alpha/2)}_{R_t^{(j)}}. \tag{4.12}$$

We observe here that for a more general nonhomogeneous half-space problem such a strong estimate of the pressure is impossible without a certain loss of regularity with respect to t.

We now turn to the estimate of the function MB (4.4) consisting of four terms. The expression

$$M_1B = -\sigma(\xi)\tilde{n}_0(\xi)\cdot\sum_{j=1}^{N_\lambda} [\Delta(\eta_j\tilde{v}_j) - \eta_j\Delta\tilde{v}_j]$$

contains no second derivates of $\tilde{v}_j$, and since $\sigma\tilde{n}_0 \in C^\alpha(\Gamma)$ it satisfies the inequality

$$\begin{aligned}|M_1B|^{(\alpha,\alpha/2)}_{S_t} &\le C_{12}(\lambda)\max_j \left[|\tilde{v}_j|^{(\alpha,\alpha/2)}_{S_t} + \sum_{m=1}^{3}\left|\frac{\partial\tilde{v}_j}{\partial\xi_m}\right|^{(\alpha,\alpha/2)}_{S_t}\right] \\ &\le \varepsilon_5\max_j |\tilde{v}_j|^{(2+\alpha,1+\alpha/2)}_{Q_t} + C_{13}(\lambda,\varepsilon_5)\max_j \int_0^t |\tilde{v}_j|^{(2+\alpha,1+\alpha/2)} d\tau\end{aligned} \tag{4.13}$$

with arbitrary $\varepsilon_5 \in (0,1)$, $t \le T$ and with a constant $C_{13}$ independent of t.

Consider

$$M_2B = -\sigma\sum_j \eta_j[\underset{\sim}{n}_0(\xi)\cdot\Delta\underset{\sim}{v}_j - \underset{\sim}{n}_0(\xi_j)\cdot\Delta_j\underset{\sim}{v}_j].$$

As we have already seen, leading coefficients of the differential operator

$$\underset{\sim}{n}_0(\xi)\Delta - \underset{\sim}{n}_0(\xi_j)\Delta_j = (\underset{\sim}{n}_0(\xi) - \underset{\sim}{n}_0(\xi_j))\Delta + \underset{\sim}{n}_0(\xi_j)(\Delta - \Delta_j)$$

are linear combinations of terms proportional to first derivatives of $\varphi_j$, and all the coefficients belong to $C^\alpha$. Therefore

$$|M_2B|_{S_t}^{(\alpha,\alpha/2)} \le (C_{14}\lambda + \varepsilon_6)\max_j |\underset{\sim}{v}_j|_{Q_t}^{(2+\alpha,1+\alpha/2)} + \tag{4.14}$$
$$+ C_{15}(\lambda,\varepsilon_6)\max_j \int_0^t |\underset{\sim}{v}_j|_{Q_\tau}^{(2+\alpha,1+\alpha/2)}d\tau.$$

The same type of estimate (but with $\lambda^\alpha$ instead of $\lambda$, since

$$|\sigma(\xi) - \sigma(\xi_j)| \le \langle\sigma\rangle^{(\alpha)}|\xi - \xi_j|^\alpha)$$

holds for

$$M_3B = -\sum_{j=1}^{N_\lambda} \eta_j[\sigma(\xi) - \sigma(\xi_j)]\underset{\sim}{n}_0(\xi_j)\cdot\Delta_j\underset{\sim}{v}_j$$

so, taking account of (4.8) we see that

$$\sum_{i=1}^{3} |M_iB|_{S_t}^{(\alpha,\alpha/2)} \le (C_{16}\lambda^\alpha + \varepsilon_7)|B|_{S_t}^{(\alpha,\alpha/2)} + C_{17}(\lambda,\varepsilon_7)\int_0^t |B|_{S_\tau}^{(\alpha,\alpha/2)}d\tau.$$

It remains to estimate the solution $(\underset{\sim}{v}'',p'')$ of the problem (4.3). First of all, the estimates of the solution of the half-space problem (4.2) which are given above make it possible to prove, by repetition of above arguments, the inequality

$$|\underset{\sim}{f}|_{Q_t}^{(\alpha,\alpha/2)} + |\rho|_{Q_t}^{(1+\alpha,\frac{1+\alpha}{2})} + |\underset{\sim}{d}|_{Q_t}^{(1+\alpha,\frac{1+\alpha}{2})} \le (C_{18}\lambda + \varepsilon_8)|B|_{S_t}^{(\alpha,\alpha/2)} \tag{4.15}$$
$$+ C_{19}(\lambda,\varepsilon_8)\int_0^t |B|_{S_\tau}^{(\alpha,\alpha/2)}d\tau, \quad \forall\varepsilon_8 \in (0,1)$$

Now, we have to verify the condition iii) for the problem (4.3). Writing $\rho$ in the form

$$\rho = \sum_{j=1}^{N_\lambda} (\nabla_j - \nabla)\cdot\eta_j\,\underset{\sim}{v} - \sum_{j=1}^{N_\lambda} \nabla_j\eta_j\cdot\underset{\sim}{v}_j$$

we obtain

$$\frac{\partial\rho}{\partial t} = \nabla\cdot\sum_{j=1}^{N_\lambda} \eta_j\left(\mathcal{J}_j^{-1} - I\right)\underset{\sim}{v}_{jt} - \sum_{j=1}^{N_\lambda} \nabla_j\eta_j\cdot(\nu\nabla_j^2\underset{\sim}{v}_j - \nabla_j p_j)$$

Since

$$-\nabla_j\eta_j\cdot(\nu\nabla_j^2\underset{\sim}{v}_j - \nabla_j p_j) = \nabla_j\cdot\left[p_j\nabla_j\eta_j - \nu\sum_{k=1}^{3}(\nabla_j\eta_j)_k\nabla_j v_{jk}\right]$$

$$- p_j\nabla_j^2\eta_j + \nu\sum_{k=1}^{3}\nabla_j(\nabla_j\eta_j)_k\cdot\nabla_j v_{jk}$$

it follows that

$$\frac{\partial\rho}{\partial t} - \nabla\cdot\underset{\sim}{f} = \nabla\cdot\underset{\sim}{h}$$

with

$$\underset{\sim}{h} = -\underset{\sim}{f} + \sum_{j=1}^{N_\lambda} \eta_j\left(\mathcal{J}_j^{-1} - I\right)\underset{\sim}{v}_{jt} + \sum_{j=1}^{N_\lambda} \mathcal{J}_j^{-1}\Big(p_j\nabla_j\eta_j$$

$$- \nu\sum_{k=1}^{3}(\nabla_j\eta_j)_k\nabla_j v_{jk}\Big) + \frac{1}{4\pi}\nabla\int_\Omega \sum_j\left(p_j\nabla_j^2\eta_j - \nu\sum_{k=1}^{3}\nabla_j(\nabla_j\eta_j)_k\cdot\nabla_j v_{jk}\right)\frac{dz}{|\xi - z|}\,.$$

In virtue of lemma 2.4 and of the above estimates for $\underset{\sim}{w}_j$ and $s_j$, we have

$$|\underset{\sim}{h}|_{Q_t}^{(\alpha,\alpha/2)} \le (C_{20}\lambda + \varepsilon_9)|B|_{S_t}^{(\alpha,\alpha/2)} + C_{21}(\lambda,\varepsilon_9)\int_0^t |B|_{S_\tau}^{(\alpha,\alpha/2)}d\tau.$$

Finally, we represent $\underset{\sim}{h}$ in a divergence form. We have

$$\underset{\sim}{h} = \underset{\sim}{h}' + \underset{\sim}{h}'' + \frac{1}{4\pi}\nabla\int_\Omega \sum_j\left(p_j\nabla_j^2\eta_j - \nu\sum_{k=1}\nabla_j(\nabla_j\eta_j)_k\cdot\nabla_j v_{jk}\right)\frac{dz}{|\xi - z|}$$

where

$$\underset{\sim}{h}' = \sum_{j=1}^{N_\lambda} \eta_j\left(\mathcal{J}_j^{-1} - I\right)\underset{\sim}{v}_{jt} - \sum_{j=1}^{N_\lambda} \eta_j\left[\nu(\nabla^2 - \nabla_j^2)\underset{\sim}{v}_j + (\nabla_j - \nabla)p_j\right]$$

$$= \sum_{j=1}^{N_\lambda} \eta_j\left(\mathcal{J}_j^{-1} - I\right)(\nu\nabla_j^2\underset{\sim}{v}_j - \nabla_j p_j) - \sum_{j=1}^{N_\lambda} \eta_j\left[\nu(\nabla^2 - \nabla_j^2)\underset{\sim}{v}_j + (\nabla_j - \nabla)p_j\right]$$

is the sum of all the terms in $\underset{\sim}{h}$ containing highest order derivatives of $\underset{\sim}{v}_j$ and $p_j$, and

$$\underset{\sim}{h}'' = \sum_{j=1}^{N_\lambda} \left[ 2\nu(\nabla\eta_j \cdot \nabla)\underset{\sim}{v}_j + \nu \underset{\sim}{v}_j \nabla^2 \eta_j - p_j \nabla\eta_j \right]$$

$$+ \sum_{j=1}^{N_\lambda} \mathcal{J}_j^{-1}\left( p_j \nabla_j \eta_j - \nu \sum_{k=1}^{3} (\nabla_j \eta_j)_k \nabla_j v_{jk} \right).$$

It is easily seen that

$$\underset{\sim}{h}' = \sum_{m=1}^{3} \frac{\partial \underset{\sim}{h}'_m}{\partial \xi_m} + \underset{\sim}{h}'''$$

with

$$\underset{\sim}{h}'_m = \sum_{j=1}^{N_\lambda} \eta_j \left( \mathcal{J}_j^{-1} - I \right) \left( \sum_{k,l=1}^{3} \nu J_j^{mk} J_j^{lk} \frac{\partial \underset{\sim}{v}_j}{\partial \xi_l} - \sum_{k=1}^{3} J_j^{mk} e_k p_j \right)$$

$$+ \sum_{j=1}^{N_\lambda} \eta_j \left[ \sum_{l=1}^{3} \nu \left( \sum_{k=1}^{3} J_j^{mk} J_j^{lk} - \delta_{ml} \right) \frac{\partial \underset{\sim}{v}_j}{\partial \xi_l} - \sum_{k=1}^{3} \left( J_j^{mk} - \delta_{mk} \right) \underset{\sim}{e}_k p_j \right],$$

$$h''' = \sum_{j=1}^{N_\lambda} \sum_{m=1}^{3} \frac{\partial}{\partial \xi_m} \left[ \eta_j \left( I - \mathcal{J}_j^{-1} \right) \right] \left( \sum_{k,l=1}^{3} \nu J_j^{mk} J_j^{lk} \frac{\partial \underset{\sim}{v}_j}{\partial \xi_l} - \sum_{k=1}^{3} J_j^{mk} \underset{\sim}{e}_k p_j \right)$$

$$- \sum_{j=1}^{N_\lambda} \frac{\partial \eta_j}{\partial \xi_m} \left[ \sum_{l=1}^{3} \nu \left( \sum_{k=1}^{3} J_j^{mk} J_j^{lk} - \delta_{ml} \right) \frac{\partial \underset{\sim}{v}_j}{\partial \xi_l} - \sum_{k=1}^{3} \left( J_j^{mk} - \delta_{mk} \right) \underset{\sim}{e}_k p_j \right].$$

Since $\underset{\sim}{h}''$ and $\underset{\sim}{h}'''$ contain no highest order derivatives of $\underset{\sim}{v}_j$ and $p_j$, it is convenient to represent these vector fields in a divergence form by means of Newtonian potentials, and we obtain finally

$$\underset{\sim}{h} = \sum_{j=1}^{3} \frac{\partial \underset{\sim}{h}_m}{\partial \xi_m},$$

$$\underset{\sim}{h}_m = \underset{\sim}{h}'_m - \frac{1}{4\pi} \frac{\partial}{\partial \xi_m} \int_\Omega \frac{\underset{\sim}{h}''(y,t) + \underset{\sim}{h}'''(y,t)}{|\xi - y|} dy + \frac{1}{4\pi} \underset{\sim}{e}_m \int_\Omega H(y,t) \frac{dy}{|\xi - y|},$$

$$H = \sum_j \left( p_j \nabla_j^2 \eta_j - \nu \sum_{k=1}^{3} \nabla_j (\nabla_j \eta_j)_k \cdot \nabla_j v_{jk} \right).$$

Clearly, for arbitrary $\gamma \in (0,1)$

$$|\underset{\sim}{h}_m|_{Q_T}^{(1+\alpha,\gamma)} \le |\underset{\sim}{h}_m'|_{Q_T}^{(1+\alpha,\gamma)} + C_{22}\left( \langle \underset{\sim}{h}'' \rangle_{t,Q_T}^{\left(\frac{1+\alpha-\gamma}{2}\right)} + \langle \underset{\sim}{h}''' \rangle_{t,Q_T}^{\left(\frac{1+\alpha-\gamma}{2}\right)} \right.$$

$$\left. + \langle H \rangle_{t,Q}^{\left(\frac{1+\alpha-\gamma}{2}\right)} \right) \le (C_{23}\lambda + \varepsilon_{10}) |B|_{S_t}^{(\alpha,\alpha/2)} + C_{24}(\lambda,\varepsilon_{10}) \int_0^t |B|_{S_\tau}^{(\alpha,\alpha/2)} d\tau.$$

The application of the theorem 1.1' to the problem (4.3) yields

$$|\underset{\sim}{v}''|_{Q_t}^{(2+\alpha,1+\alpha/2)} \le (C_{25}\lambda + \varepsilon_{11}) |B|_{S_t}^{(\alpha,\alpha/2)} + C_{25}(\lambda,\varepsilon_{10}) \int_0^t |B|_{S_\tau}^{(\alpha,\alpha/2)} d\tau,$$

hence, the same type of estimate holds for $M_4 B = -\sigma \underset{\sim}{n}_0 \cdot \Delta \underset{\sim}{v}''$. This completes the proof of (4.5) and the theorem 1.1.

At the conclusion we remark that we could give another proof of our existence theorem, independent of theorem 1.1', which had been used only for the sake of convenience.

### §5. Proof of theorem 1.2.

As in the case of theorem 1.1, we begin with auxiliary constructions. We write problem (1.7) in the form

$$\underset{\sim}{w}_t - \nu\nabla^2 \underset{\sim}{w} + \nabla s = \underset{\sim}{f} + \underset{\sim}{l}_1(\underset{\sim}{w},s) = \underset{\sim}{f}_1,$$

$$\nabla \cdot \underset{\sim}{w} = \rho + l_2(\underset{\sim}{w}) \equiv \rho_1,$$

$$\nu \Pi_0 S(\underset{\sim}{w}) \underset{\sim}{n}_0 |_{\xi \in \Gamma} = \underset{\sim}{l}_3(\underset{\sim}{w}) + \Pi_0 \underset{\sim}{d} \qquad (5.1)$$

$$\underset{\sim}{n}_0 \cdot T(\underset{\sim}{w},s) \underset{\sim}{n}_0 - \sigma \underset{\sim}{n}_0 \cdot \Delta(0) \int_0^t \underset{\sim}{w} d\tau |_{\xi \in \Gamma} = b + \int_0^t B d\tau + l_4(\underset{\sim}{w},s) + \int_0^t l_5(\underset{\sim}{w}) d\tau$$

where

$$\underset{\sim}{l}_1(\underset{\sim}{w}, s) = \nu(\nabla_u^2 - \nabla^2)\underset{\sim}{w} + (\nabla - \nabla_u)s$$

$$= \sum_{j,m=1}^{3} \frac{\partial}{\partial \xi_j}\left(\sum_{l=1}^{3} A_{lj}A_{lm} - \delta_{jm}\right)\frac{\partial \underset{\sim}{w}}{\partial \xi_m} + (I - \mathcal{A})\nabla s,$$

$$l_2(\underset{\sim}{w}) = (\nabla - \nabla_u)\cdot \underset{\sim}{w},$$

$$\underset{\sim}{l}_3(\underset{\sim}{w}) = \nu(\Pi_0 S(\underset{\sim}{w})\underset{\sim}{n}_0 - \Pi_0 \Pi S_u(\underset{\sim}{w})\underset{\sim}{n}), \tag{5.2}$$

$$l_4(\underset{\sim}{w}, s) = \underset{\sim}{n}_0\cdot(T\underset{\sim}{n}_0 - T_u\underset{\sim}{n}) = s(\underset{\sim}{n} - \underset{\sim}{n}_0\cdot\underset{\sim}{n}_0) + \nu(\underset{\sim}{n}_0\cdot(S(\underset{\sim}{w})\underset{\sim}{n}_0 - S_u(\underset{\sim}{w})\underset{\sim}{n})),$$

$$l_5(\underset{\sim}{w}) = \sigma\underset{\sim}{n}_0\cdot\frac{\partial}{\partial t}\left[(\Delta(t) - \Delta(0))\int_0^t \underset{\sim}{w}d\tau\right]$$

$$= \sigma\underset{\sim}{n}_0\cdot\left[(\Delta(t) - \Delta(0))\underset{\sim}{w} + \dot{\Delta}(t)\int_0^t \underset{\sim}{w}d\tau\right] \equiv l_{15} + l_{25}$$

and $\dot{\Delta}(t)$ denotes a differential operator on $\Gamma$ whose coefficients are derivatives of coefficients of $\Delta(t)$ with respect to t.

We recall that $A_{ij}$ are cofactors of elements of the Jacobi matrix of the transformation $X_u$ and that

$$\sum_{m=1}^{3} \frac{\partial}{\partial \xi_m} A_{mj} = 0 \ (j = 1,2,3)$$

which makes it possible to represent $\underset{\sim}{l}_1(\underset{\sim}{w}, s)$ and $l_2(\underset{\sim}{w})$ in the divergence form according to the formulas

$$\underset{\sim}{l}_1(\underset{\sim}{w}, s) = \sum_{m=1}^{3} \frac{\partial}{\partial \xi_j}\underset{\sim}{L}_{1j}(\underset{\sim}{w}, s), \quad \underset{\sim}{l}_2(\underset{\sim}{w}) = \nabla\cdot\underset{\sim}{L}_2(\underset{\sim}{w})$$

with

$$\begin{aligned} \underset{\sim}{L}_{1j}(\underset{\sim}{w}, s) &= \sum_{m=1}^{3}\left(\sum_{l=1}^{3} A_{lj}A_{lm} - \delta_{jm}\right)\frac{\partial \underset{\sim}{w}}{\partial \xi_m} + (I - \mathcal{A})\underset{\sim}{e}_j s \\ &= \sum_{m=1}^{3}\left(B_{lj}A_{lj} + B_{jm}\right)\frac{\partial \underset{\sim}{w}}{\partial \xi_m} - \mathcal{B}\underset{\sim}{e}_j s, \end{aligned} \tag{5.3}$$

$$\underset{\sim}{L}_2(\underset{\sim}{w}) = (I - \mathcal{A}^*)\underset{\sim}{w} \equiv -\mathcal{B}^*\underset{\sim}{w}$$

where $\mathcal{B}$ is a matrix with the elements $B_{ij} = A_{ij} - \delta_{ij}$, i.e. $\mathcal{B} = \mathcal{A} - I$.

Let us write the formulas (1.9′) for $\rho_1$ and $\underset{\sim}{f}_1$ with appropriate $\underset{\sim}{h}$ and $\underset{\sim}{h}_m$ assuming that the condition iii) holds for $\rho$ and $\underset{\sim}{f}$ with a given $\underset{\sim}{u}$. We have

$$\frac{\partial\rho_1}{\partial t} - \nabla\cdot\underset{\sim}{f}_1 = \frac{\partial\rho}{\partial t} - \nabla\cdot\underset{\sim}{f} + \frac{\partial l_2(\underset{\sim}{w})}{\partial t} - \nabla\cdot\underset{\sim}{l}_1(\underset{\sim}{w},s)$$

$$= \frac{\partial\rho}{\partial t} - \nabla\cdot\underset{\sim}{f} + \nabla\cdot[(I - \mathcal{A}^*)\underset{\sim}{w}_t - \mathcal{A}^*_t\underset{\sim}{w} - \underset{\sim}{l}_1(\underset{\sim}{w},s)] = \frac{\partial\rho}{\partial t} - \nabla_u\cdot\underset{\sim}{f}$$

$$+ \nabla\cdot[(I - \mathcal{A}^*)(\underset{\sim}{w}_t - \underset{\sim}{f}) - \mathcal{A}^*_t\underset{\sim}{w} - \underset{\sim}{l}_1] = \nabla\cdot(\underset{\sim}{h} + \underset{\sim}{l}_6(\underset{\sim}{w},s))$$

where

$$\underset{\sim}{l}_6(\underset{\sim}{w},s) = -\mathcal{B}^*(\nu\nabla^2\underset{\sim}{w} - \nabla s) - \mathcal{B}^*_t\underset{\sim}{w} - \mathcal{A}^*\underset{\sim}{l}_1 \tag{5.4}$$

This expression can be written in the following divergence form

$$\underset{\sim}{l}_6(\underset{\sim}{w},s) = \sum_{j=1}^{3}\frac{\partial \underset{\sim}{L}_{6j}}{\partial\xi_j},$$

$$\underset{\sim}{L}_{6j}(\underset{\sim}{w},s) = -\nu\mathcal{B}^*\frac{\partial\underset{\sim}{w}}{\partial\xi_j} + \mathcal{B}^* e_j s - \mathcal{A}^* L_{1j} + \frac{\partial V}{\partial\xi_j}, \tag{5.5}$$

$$\underset{\sim}{V} = -\frac{1}{4\pi}\int_\Omega\frac{1}{|\xi - \eta|}\left[\sum_{m=1}^{3}\left(\nu\frac{\partial\mathcal{B}^*}{\partial\eta_m}\frac{\partial\underset{\sim}{w}}{\partial\eta_m} - \frac{\partial\mathcal{B}^*}{\partial\eta_m}\underset{\sim}{e}_m s + \frac{\partial\mathcal{B}^*}{\partial\eta_m}\underset{\sim}{L}_{1m}\right) - \mathcal{B}^*_t\underset{\sim}{w}\right]d\eta.$$

As seen from the definition (5.2), the coefficients of differential operators $\underset{\sim}{l}_1$, $l_2$, $\underset{\sim}{L}_{1j}$, $L_2$, $\underset{\sim}{l}_6$, $\underset{\sim}{L}_{6j}$ depend on $B_{ij}$ and sometimes on the derivatives of $B_{ij}$. Furthermore, as

$$\underset{\sim}{n} - \underset{\sim}{n}_0 = \frac{\mathcal{A}\underset{\sim}{n}_0}{|\mathcal{A}\underset{\sim}{n}_0|} - \underset{\sim}{n}_0 = \frac{\mathcal{B}\underset{\sim}{n}_0}{|\mathcal{A}\underset{\sim}{n}_0|} - \underset{\sim}{n}_0\frac{\mathcal{A}\underset{\sim}{n}_0\cdot\mathcal{A}\underset{\sim}{n}_0 - \underset{\sim}{n}_0\cdot\underset{\sim}{n}_0}{|\mathcal{A}\underset{\sim}{n}_0|(|\mathcal{A}\underset{\sim}{n}_0| + 1)},$$

the coefficient of

$$\underset{\sim}{l}_3(\underset{\sim}{w}) = \nu\Big[\Pi_0(\Pi_0 - \Pi)S(\underset{\sim}{w})\underset{\sim}{n}_0 + \Pi_0\Pi(S(\underset{\sim}{w}) - S_u(\underset{\sim}{w}))\underset{\sim}{n}_0 + \Pi_0\Pi S_u(\underset{\sim}{w})(\underset{\sim}{n}_0 - \underset{\sim}{n})\Big]$$

and

$$l_4(\underset{\sim}{w},s) = s(\underset{\sim}{n} - \underset{\sim}{n}_0)\cdot\underset{\sim}{n}_0 + \nu\underset{\sim}{n}_0\cdot\Big[(S(\underset{\sim}{w}) - S_u(\underset{\sim}{w}))\underset{\sim}{n}_0 + S_u(\underset{\sim}{w})(\underset{\sim}{n}_0 - \underset{\sim}{n})\Big]$$

also depend on $B_{ij}$, in fact, they are linear combinations of functions proportional to $B_{ij}$. Finally, we write operators $\Delta(t)$ and $\mathring{\Delta}(t)$ in the neighbourhood of the point $\xi_0$, taking the coordinates on the tangent plane to $\Gamma$ at the point $\xi_0$ as local coordinates on $\Gamma$ and on $\Gamma_t = X_u\Gamma$. To simplify the calculations we shall assume that $\xi_0 = 0$ and that the coordinate system $(\xi_1, \xi_2, \xi_3)$ coincides with a local cartesian system $(y_1, y_2, y_3)$ at the point $\xi_0$. Then $(\xi_1,\xi_2)$ will be local coordinates on $\Gamma$ and on $\Gamma_t$ near the points $\xi_0 = 0$ and $X_u0$, respectively. The surface $\Gamma$ will be given by equation (2.1) i.e.

$$\xi_3 = \varphi(\xi'), \qquad \xi' \in K_d,$$

and the points of $\Gamma_t$ will have the cartesian coordinates $\xi_1(\xi',t)$ equal to

$$\xi_1 + \int_0^t \omega_1(\xi',\tau)d\tau,\ \xi_2 + \int_0^t \omega_2(\xi',\tau)d\tau,\ \varphi(\xi') + \int_0^t \omega_3(\xi',\tau)d\tau$$

where $\omega_1(\xi',t) = u_1(\xi_1,\xi_2,\varphi(\xi'),t)$. The elements of the metric tensor of $\Gamma$ will be defined by

$$g_{\alpha\beta}(\xi',t) = \sum_{1=1}^{3} \frac{\partial\xi_1(\xi',t)}{\partial\xi_\alpha}\,\frac{\partial\xi_1(\xi',t)}{\partial\xi_\beta} = \delta_{\alpha\beta} + \varphi_\alpha\varphi_\beta \tag{5.6}$$
$$+ \varphi_\alpha d_{3\beta} + \varphi_\beta d_{3\alpha} + d_{\alpha\beta} + d_{\beta\alpha} + \sum_{1=1}^{3} d_{1\alpha}d_{1\beta}$$

where

$$d_{1\alpha} = \int_0^t \frac{\partial\omega_1}{\partial\xi_\alpha}\,d\tau = \int_0^t \left(\frac{\partial u_1}{\partial\xi_\alpha} + \varphi_\alpha\frac{\partial u_1}{\partial\xi_3}\right)\Big|_{\xi_3=\varphi(\xi')} d\tau \tag{5.7}$$

and $\varphi_\alpha = \dfrac{\partial\varphi}{\partial\xi_\alpha}$.

Let us write the Laplace-Beltrami operator in the form

$$\Delta(t) = \sum_{\alpha,\beta=1}^{2} g^{\alpha\beta}\frac{\partial^2}{\partial\xi_\alpha\partial\xi_\beta} + \sum_{\beta=1}^{2} h_\beta\frac{\partial}{\partial\xi_\beta}$$

where

$$h_\beta = \frac{1}{\sqrt{g}}\sum_{\alpha=1}^{2}\frac{\partial}{\partial\xi_\alpha}g^{\alpha\beta}\sqrt{g}\,.$$

Clearly,

$$\Delta(t) - \Delta(0) = \sum_{\alpha,\beta=1}^{2} \left(g^{\alpha\beta}(\xi',t) - g^{\alpha\beta}(\xi',0)\right)\frac{\partial^2}{\partial\xi_\alpha\partial\xi_\beta}$$

$$+ \sum_{\beta=1}^{2} \left(h_\beta(\xi',t) - h_\beta(\xi',0)\right)\frac{\partial}{\partial\xi_\beta}$$

and

$$\dot{\Delta}(t) = \sum_{\alpha,\beta=1}^{2} \frac{\partial g^{\alpha\beta}(\xi',t)}{\partial t}\frac{\partial^2}{\partial\xi_\alpha\partial\xi_\beta} + \sum_{\beta=1}^{2} \frac{\partial h_\beta(\xi',t)}{\partial t}\frac{\partial}{\partial\xi_\alpha} .$$

As seen from (5.6),

$$g_{\alpha\beta}(\xi',t) - g_{\alpha\beta}(\xi',0) = \varphi_\alpha d_{3\beta} + \varphi_\beta d_{3\alpha} + d_{\alpha\beta} + d_{\beta\alpha} + \sum_{i=1}^{3} d_{i\alpha}d_{i\beta}$$

is a sum of terms containing $d_{11}$ or $d_{12}$ as a factor. Similar structure have

$$g^{\alpha\beta}(\xi' t) - g^{\alpha\beta}(\xi' 0) \text{ and } h_\beta(\xi',t) - h_\beta(\xi',0),$$

but the latter difference contains also terms proportional to the first derivatives of $d_{i\alpha}$. Coefficients of $\dot{\Delta}(t)$ are sums of terms proportional to

$$\frac{\partial\omega_i}{\partial\xi_\alpha} \text{ or to } \frac{\partial^2\omega_i}{\partial\xi_\alpha\partial\xi_\beta} .$$

We now turn to estimates of $l_i$ for which purpose we first of all consider the functions $B_{ij} = A_{ij} - \delta_{ij}$. It follows from the definition of $A_{ij}$ that

$$B_{jj} = b_{jj} + b_{kk} + b_{jj}b_{kk} - b_{jk}b_{kj},$$
$$B_{ij} = -b_{ji} + b_{ki}b_{kj} - b_{ji}b_{kk}, \qquad (5.8)$$

where $i \neq j$, $j \neq k$, $k \neq i$, and

$$b_{ij}(\xi,t) = \int_0^t \frac{\partial u_i(\xi,\tau)}{\partial\xi_j}\, d\tau.$$

**Lemma 5.1** *Suppose that* $\underset{\sim}{u}(\xi,t)$ *satisfies conditions (1.11) with* $\delta < 1$. *Then*

$$\sup_{t\le T}|B_{ij}|_\Omega^{(\alpha)} \le c_1 \max_{k,m} \sup_{t\le T} |b_{km}|_\Omega^{(\alpha)}, \qquad (5.9_1)$$

$$\sup_{t\le T}|B_{ij}|_{\Omega}^{(1+\alpha)} \le c_2\max_{k,m}\sup|b_{km}|_{\Omega}^{(1+\alpha)}, \qquad (5.9_2)$$

$$\langle B_{ij}\rangle_{t,Q_T}^{(\beta)} \le c_3\max_{k,m}\langle b_{km}\rangle_{t,Q_T}^{(\beta)}, \quad 0<\beta<1, \qquad (5.9_3)$$

$$|B_{ij}|_{Q_T}^{(1+\alpha,\gamma)} \le c_4\max_{k,m}|b_{km}|_{Q_T}^{(1+\alpha,\gamma)}, \qquad (5.9_4)$$

$$\left|\frac{\partial B_{ij}}{\partial t}\right|_{Q_T}^{(\alpha,\alpha/2)} \le c_5\max_{k,m}\left|\frac{\partial b_{km}}{\partial t}\right|_{Q_T}^{(\alpha,\alpha/2)} \le c_5|Du|_{Q_T}^{(\alpha,\alpha/2)} \qquad (5.9_5)$$

$$\left\langle\frac{\partial B_{ij}}{\partial t}\right\rangle_{t,Q_T}^{\left(\frac{1+\alpha-\gamma}{2}\right)} \le c_6\left(\max_{k,m}\left\langle\frac{\partial b_{km}}{\partial t}\right\rangle_{t,Q_T}^{\left(\frac{1+\alpha-\gamma}{2}\right)} + \max_{k,m}\left|\frac{\partial b_{km}}{\partial t}\right|_{Q_T}\right)$$

$$\le c_6\left(\langle Du\rangle_{t,Q_T}^{\left(\frac{1+\alpha-\gamma}{2}\right)} + |Du|_{Q_T}\right), \qquad (5.9_6)$$

$$\left|\frac{\partial B_{ij}}{\partial t}\right|_{Q_T}^{(1+\alpha,\gamma)} \le c_6\left(\max_{k,m}\left|\frac{\partial b_{km}}{\partial t}\right|_{Q_T}^{(1+\alpha,\gamma)} + \max_{k,m}\sup_{t\le T}\left|\frac{\partial b_{km}}{\partial t}\right|^{(\gamma)}\right)$$

$$\le c_6\left(|Du|_{Q_T}^{(1+\alpha,\gamma)} + \sup_{t\le T}|Du|_{\Omega}^{(\gamma)}\right) \qquad (5.9_7)$$

Proof. The proof of $(5.9_1)$ - $(5.9_4)$ reduces to the application of inequalities (2.22), (2.23) and of the estimates

$$|b_{km}|_{\Omega}^{(1+\alpha)} \le \int_0^t \left|\frac{\partial u_k}{\partial \xi_m}\right|_{\Omega}^{(1+\alpha)} d\tau \le \delta, \qquad (5.10_1)$$

$$|b_{km}|_{Q_T}^{(1+\alpha,\gamma)} \le T^{\frac{(1-\alpha+\gamma)}{2}}\sup_{t\le T}\left|\frac{\partial u_k}{\partial \xi_m}\right|_{\Omega}^{(\gamma)} \le \delta. \qquad (5.10_2)$$

For instance, in virtue of (2.23) and (5.8), we have

$$|B_{ij}|_{Q_T}^{(1+\alpha,\gamma)} \le \max_{k,m}|b_{km}|_{Q_T}^{(1+\alpha,\gamma)}(2 + 8\max_{l,n}|\sup_{t\le T} b_{ln}|_{\Omega}^{(\gamma)})$$

$$\le (2+8\delta)\max_{k,m}|b_{km}|_{Q_T}^{(1+\alpha,\gamma)}.$$

The estimates $(5.9_5)$ - $(5.9_7)$ are proved by the same arguments.

Inequalities $(5.9_2)$, $(5.9_4)$, (5.10) have an important corollary, namely

$$\sup_{t\le T} |B_{ij}(\cdot,t)|_{\Omega}^{(1+\alpha)} + |B_{ij}|_{Q_T}^{(1+\alpha,\gamma)} \le c_8\delta \tag{5.11}$$

which is a basis of the following proposition.

**Lemma 5.2** *Suppose that* $u(\xi,t)$ *satisfies the conditions (1.11 ) and that* $\Gamma \in C^{2+\alpha}$. *Then*

$$\begin{aligned}
&|\underset{\sim}{l}_1(\underset{\sim}{w},s)|_{Q_T}^{(\alpha,\alpha/2)} + |\nabla l_2(\underset{\sim}{w})|_{Q_T}^{(\alpha,\alpha/2)} + |l_2(\underset{\sim}{w})|_{Q_T}^{(1+\alpha,\gamma)} \\
&+ \sup_{t\le T} |l_4(\underset{\sim}{w},s)|_{\Gamma}^{(1+\alpha)} + |l_5(\underset{\sim}{w})|_{S_T}^{(\alpha,\alpha/2} + \sup_{t\le T} |\underset{\sim}{l}_3(\underset{\sim}{w},s)|_{\Gamma}^{(1+\alpha)} \\
&\le c_9\delta\left(|\underset{\sim}{w}|_{Q_T}^{(2+\alpha,1+\alpha/2)} + |\nabla s|_{Q_T}^{(\alpha,\alpha/2)} + \sup_{t\le T} |s(\cdot,t)|_{\Gamma}^{(1+\alpha)}\right),
\end{aligned} \tag{5.12}$$

$$\begin{aligned}
|L_{1k}(\underset{\sim}{w},s)|_{Q_T}^{(1+\alpha,\gamma)} &\le c_{10}\delta\left(|\underset{\sim}{w}|_{Q_T}^{(2+\alpha,1+\alpha/2)} + \sup_{t\le T} |s(\cdot,t)|_{\Omega}^{(\gamma} \right. \\
&\left. + |s|_{Q_T}^{(1+\alpha,\gamma)}\right)
\end{aligned} \tag{5.13}$$

$$\begin{aligned}
&\langle l_2(\underset{\sim}{w})\rangle_{t,S_T}^{(\frac{1+\alpha}{2})} + \langle \underset{\sim}{l}_3(\underset{\sim}{w})\rangle_{t,S_T}^{(\frac{1+\alpha}{2})} + \langle l_4(\underset{\sim}{w},s)\rangle_{t,S_T}^{(\frac{1+\alpha}{2})} \\
&\le c_{11}\delta\left(|\underset{\sim}{w}|_{Q_T}^{(2+\alpha,1+\alpha/2)} + \langle s\rangle_{t,S_T}^{(\frac{1+\alpha}{2})}\right) + c_{12}T^{\frac{1-\alpha}{2}} |D\underset{\sim}{u}|_{S_T}\left(|D\underset{\sim}{w}|_{S_T} + |s|_{S_T}\right)
\end{aligned} \tag{5.14}$$

Proof All the terms in the left side of (5.12) are estimated in a similar way, for example, in virtue of inequalities (2.26) and of (5.11),

$$\begin{aligned}
|\underset{\sim}{l}_1(\underset{\sim}{w},s)|_{Q_T}^{(\alpha,\alpha/2)} &\le c_{13}K_1\left(|\underset{\sim}{w}|_{Q_T}^{(2+\alpha,1+\alpha/2)} + |\nabla s|_{Q_T}^{(\alpha,\alpha/2)}\right) \\
&\le c_{14}\delta\left(|\underset{\sim}{w}|_{Q_T}^{(2+\alpha,1+\alpha/2)} + |\nabla s|_{Q_T}^{(\alpha,\alpha/2)}\right)
\end{aligned}$$

where

$$K_1 = \max_{i,j} |B_{ij}|^{(\alpha,\alpha/2)}_{Q_T} + \max_{i,j} |\nabla B_{ij}|^{(\alpha,\alpha/2)}_{Q_T}.$$

As has been already pointed out, norms of coefficients of $l_2$, $\underset{\sim}{l}_3$, $l_4$ are also estimated in terms of $K_1$. The same type of estimate holds for $l_{15}(\underset{\sim}{w}) = (\Delta(t) - \Delta(0))\underset{\sim}{w}$:

$$|l_{16}(\underset{\sim}{w})|^{(\alpha,\alpha/2)}_{S_T} \le c_{14}K_2|\underset{\sim}{w}|^{(2+\alpha,1+\alpha/2)}_{S_T} \le c_{15}\delta|\underset{\sim}{w}|^{(2+\alpha,1+\alpha/2)}_{S_T}$$

Here

$$K_2 = \sup_{\xi_0}\left(\max_{i,\beta} |d_{i\beta}|^{(\alpha,\alpha/2)}_{K_d} + \max_{i,\beta,\gamma} |\frac{\partial}{\partial y_\gamma} d_{i\beta}|^{(\alpha,\alpha/2)}_{K_d}\right)$$

where $d_{i\beta}$ is defined by (5.6) in the neighbourhood of an arbitrary fixed point $\xi_0 \in \Gamma$, and the supremum is taken over all $\xi_0 \in \Gamma$. For

$$l_{25}(\underset{\sim}{w}) = \dot{\Delta}(t)\int_0^t \underset{\sim}{w}d\tau$$

we have

$$|l_{25}(\underset{\sim}{w})|^{(\alpha,\alpha/2)}_{S_T} \le c_{16}K_3\left(|\int_0^t D\underset{\sim}{w}d\tau|^{(\alpha,\alpha/2)}_{S_T}\right.$$

$$\left.+ |\int_0^t D^2\underset{\sim}{w}d\tau|^{(\alpha,\alpha/2)}_{S_T}\right) \le c_{17}K_3(T + T^{1-\alpha/2})|\underset{\sim}{w}|^{(2+\alpha,1+\alpha/2)}_{Q_T} \le c_{18}\delta|\underset{\sim}{w}|^{(2+\alpha,1+\alpha/2)}_{Q_T}$$

where

$$K_3 = \sup_{\xi_0}\left(\max_{i,\beta}\left|\frac{\partial\omega_i}{\partial y_\beta}\right|^{(\alpha,\alpha/2)}_{K_d\times(0,T)} + \max_{i,\beta,\gamma}\left|\frac{\partial^2\omega_i}{\partial y_\beta\partial y_\gamma}\right|^{(\alpha,\alpha/2)}_{K_d\times(0,T)}\right)$$

and

$$D^2\underset{\sim}{\omega} = \left\{\frac{\partial^2\omega_i}{\partial\xi_j\partial\xi_k}\right\}_{i,j,k=1,2,3}.$$

The inequality (5.13) follows from (5.3), (2.23), $(5.10_1)$ and lemma 2.2. Finally, in virtue of $(5.9_3)$,

$$\langle l_2(\underset{\sim}{w})\rangle^{(\frac{1+\alpha}{2})}_{t,S_T} + \langle \underset{\sim}{l}_3(\underset{\sim}{w})\rangle^{(\frac{1+\alpha}{2})}_{t,S_T} + \langle l_4(\underset{\sim}{w},s)\rangle^{(\frac{1+\alpha}{2})}_{t,S_T}$$

$$\le c_{19}\left\{\max_{i,j} |B_{ij}|_{S_T}\left(\langle D\underset{\sim}{w}\rangle^{(\frac{1+\alpha}{2})}_{t,S_T} + \langle s\rangle^{(\frac{1+\alpha}{2})}_{t,S_T}\right) + \max_{i,j} \langle B_{ij}\rangle^{(\frac{1+\alpha}{2})}_{t,Q_T}\left(|D\underset{\sim}{w}|_{S_t} + |s|_{S_T}\right)\right\}$$

$$\le c_{11}\delta\left(|\underset{\sim}{w}|^{(2+\alpha,1+\alpha/2)}_{Q_T} + \langle s\rangle^{\left(\frac{1+\alpha}{2}\right)}_{t,S_T}\right) + c_{12}T^{\frac{1-\alpha}{2}}|D\underset{\sim}{u}|_{S_T}\left(|D\underset{\sim}{w}|_{S_T} + |s|_{S_T}\right),$$

and the lemma is proved.

It remains to estimate $\underset{\sim 6}{l}$ and $\underset{\sim 6j}{L}$ defined by (5.4) and (5.5).

**Lemma 5.3** *Under the hypotheses of lemma 5.2 the following estimates hold:*

$$|\underset{\sim 6}{l}(\underset{\sim}{w},s)|^{(\alpha,\alpha/2)}_{Q_T} \le c_{20}\delta\left(|\underset{\sim}{w}|^{(2+\alpha,1+\alpha/2)}_{Q_T} + |\nabla s|^{(\alpha,\alpha/2)}_{Q_T}\right)$$
$$+ c_{21}|D\underset{\sim}{u}|^{(\alpha,\alpha/2)}|\underset{\sim}{w}(\cdot,0)|^{(\alpha)}_{\Omega}, \tag{5.15}$$

$$|\underset{\sim 6m}{L}(\underset{\sim}{w},s)|^{(1+\alpha,\gamma)}_{Q_T} \le c_{22}\delta\left(|\underset{\sim}{w}|^{(2+\alpha,1+\alpha/2)}_{Q_T} + |s|^{(1+\alpha,\gamma)}_{Q_T}\right)$$
$$+ c_{23}\left(\langle D\underset{\sim}{u}\rangle^{\left(\frac{1+\alpha-\gamma}{2}\right)}_{t,Q_T}|\underset{\sim}{w}(\cdot,0)|^{(\alpha)}_{\Omega} + |D\underset{\sim}{u}|_{Q_T}|\underset{\sim}{w}(\cdot,0)|^{(\alpha)}_{\Omega}\right) \tag{5.16}$$

<u>Proof</u>. Consider the function $B^*_t\underset{\sim}{w}$. In virtue of (2.22) and of $(5.9_5)$

$$|B^*_t\underset{\sim}{w}|^{(\alpha,\alpha/2)}_{Q_T} \le c_{24}|D\underset{\sim}{u}|^{(\alpha,\alpha/2)}_{Q_T}\,|\underset{\sim}{w}|^{(\alpha,\alpha/2)}_{Q_T}$$

The elementary estimate of the norm of $\underset{\sim}{w}$, namely,

$$|\underset{\sim}{w}|^{(\alpha,\alpha/2)}_{Q_T} \le |\underset{\sim}{w}(\cdot,0)|^{(\alpha)}_{\Omega} + \int_0^T |\underset{\sim t}{w}(\cdot,t)|^{(\alpha)}_{\Omega}dt$$
$$+ |\underset{\sim t}{w}|_{Q_T}T^{1-\alpha/2} \le |\underset{\sim}{w}(\cdot,t)|^{(\alpha)}_{\Omega} + 2|\underset{\sim}{w}|^{(2+\alpha,1+\alpha/2)}_{Q_T}(T + T^{1-\alpha/2})$$

leads to the inequality

$$|B^*_t\underset{\sim}{w}|^{(\alpha,\alpha/2)}_{Q_T} \le c_{24}|D\underset{\sim}{u}|^{(\alpha,\alpha/2)}_{Q_T}|\underset{\sim}{w}(\cdot,0)|^{(\alpha)}_{\Omega} + c_{25}\delta|\underset{\sim}{w}|^{(2+\alpha,1+\alpha/2)}_{Q_T}.$$

Other terms in (5.4) are estimated precisely as $\underset{\sim 1}{l}(\underset{\sim}{w},s)$, which yields (5.15).

Let us prove inequality (5.16). In virtue of (2.23) and of $(5.10_2)$ we have

$$|\underset{\sim}{L}_{6j}|_{Q_T}^{(1+\alpha,\gamma)} \le c_{26}\delta\Big(|\underset{\sim}{w}|_{Q_T}^{(2+\alpha,1+\alpha/2)} + |s|_{Q_T}^{(1+\alpha,\gamma)}$$
$$+ \sup_{t\le T}|s(\cdot,t)|_{\Omega}^{(\gamma)}\Big) + \left|\frac{\partial V}{\partial \xi_j}\right|_{Q_T}^{(1+\alpha,\gamma)}. \tag{5.17}$$

We apply lemma 2.4. to V to obtain

$$\left|\frac{\partial V}{\partial \xi_j}\right|_{Q_T}^{(1+\alpha,\gamma)} \le c_{27}\Big\{\delta\Big(|\underset{\sim}{w}|_{Q_T}^{(2+\alpha,1+\alpha/2)} + |s|_{Q_T}^{(1+\alpha,\gamma)}$$
$$+ \sup_{t\le T}|s(\cdot,t)|_{\Omega}^{(\gamma)}\Big) + \langle B_t^*\underset{\sim}{w}\rangle_{t,Q_T}^{\left(\frac{1+\alpha-\gamma}{2}\right)}\Big\}. \tag{5.18}$$

Finally, in virtue of $(5.9_6)$

$$\langle B_t^*\underset{\sim}{w}\rangle_{t,Q_T}^{\left(\frac{1+\alpha-\gamma}{2}\right)} \le c_{28}\left(\langle D\underset{\sim}{u}\rangle_{t,Q_T}^{\left(\frac{1+\alpha-\gamma}{2}\right)} + |D\underset{\sim}{u}|_{Q_T}\right)\Bigg(|\underset{\sim}{w}|_{Q_T}$$

$$+ \langle \underset{\sim}{w}\rangle_{t,Q_T}^{\left(\frac{1+\alpha-\gamma}{2}\right)}\Bigg) \le c_{28}\left(\langle D\underset{\sim}{u}\rangle_{t,Q_T}^{\left(\frac{1+\alpha-\gamma}{2}\right)} + |D\underset{\sim}{u}|_{Q_T}\right)\left(|\underset{\sim}{w}(\cdot,0)|_{\Omega}+(T + T^{\frac{1-\alpha+\gamma}{2}})|\underset{\sim}{w}_t|_{Q_T}\right)$$

$$\le c_{29}|\underset{\sim}{w}(\cdot,0)|_{\Omega}\left(\langle D\underset{\sim}{u}\rangle_{t,Q_T}^{\left(\frac{1+\alpha-\gamma}{2}\right)} + |D\underset{\sim}{u}|_{Q_T}\right) + c_{30}\delta|\underset{\sim}{w}|_{Q_T}^{(2+\alpha,1+\alpha/2)}.$$

This inequality together with (5.17) and (5.18) implies (5.16). The lemma is proved.

Let us summarize our preliminary estimates.

**Lemma 5.4** *If* $\underset{\sim}{u}$ *satisfies conditions (1.11) and* $\Gamma \in C^{2+\alpha}$, *then*

$$|\underset{\sim}{l}_1(\underset{\sim}{w},s)|_{Q_T}^{(\alpha,\alpha/2)} + |\nabla l_2(\underset{\sim}{w})|_{Q_T}^{(\alpha,\alpha/2)} + |l_2(\underset{\sim}{w})|_{Q_T}^{(1+\alpha,\gamma)} +$$
$$|\underset{\sim}{l}_6(\underset{\sim}{w},s)|_{Q_T}^{(\alpha,\alpha/2)} + \sum_{m=1}^{3}|L_{6m}(\underset{\sim}{w},s)|_{Q_T}^{(1+\alpha,\gamma)} + |l_2(\underset{\sim}{w})|_{S_T}^{(1+\alpha,\frac{1+\alpha}{2})} +$$
$$+ |\underset{\sim}{l}_3(\underset{\sim}{w})|_{S_T}^{(1+\alpha,\frac{1+\alpha}{2})} + |l_4(\underset{\sim}{w})|_{S_T}^{(1+\alpha,\frac{1+\alpha}{2})} + |l_5(\underset{\sim}{w})|_{S_T}^{(\alpha,\alpha/2)} \le \tag{5.19}$$
$$\le c_{30}\delta\Big(|\underset{\sim}{w}|_{Q_T}^{(2+\alpha,1+\alpha/2)} + |\nabla s|_{Q_T}^{(\alpha,\alpha/2)} + |s|_{S_T}^{(1+\alpha,\frac{1+\alpha}{2})} +$$
$$+ |s|_{Q_T}^{(1+\alpha,\gamma)}\Big) + c_{32}P_T[\underset{\sim}{u}]\Big(|\underset{\sim}{w}(\cdot,0)|_{\Omega} + |D\underset{\sim}{w}(\cdot,0)|_{\Omega} + |s(\cdot,0)|_{\Gamma}\Big)$$

*where* $P_T[\underset{\sim}{u}]$ *is given by (1.13).*

Proof. Add the inequalities (5.12) – (5.16) and observe that

$$|\underset{\sim}{w}(\cdot,0)|_{\Omega}^{(\alpha,)} \le c_{33}(|\underset{\sim}{w}(\cdot,0)|_{\Omega} + |D\underset{\sim}{w}(\cdot,0)|_{\Omega}),$$

$$T^{\frac{1-\alpha}{2}}|D\underset{\sim}{u}|_{S_T}\left(|D\underset{\sim}{w}|_{S_T} + |s|_T\right) \le T^{\frac{1-\alpha}{2}}|D\underset{\sim}{u}|_{S_T}\left(|D\underset{\sim}{w}(\cdot,0)|_{\Gamma}\right.$$

$$\left. + |s(\cdot,0)|_{\Gamma} + T^{\frac{1+\alpha}{2}}\langle D\underset{\sim}{w}\rangle_{t,S_T}^{\left(\frac{1+\alpha}{2}\right)} + T^{\frac{1+\alpha}{2}}\langle s\rangle_{t,Q_T}^{\left(\frac{1+\alpha}{2}\right)}\right)$$

$$\le \delta\left(|\underset{\sim}{w}|_{Q_T}^{(2+\alpha,1+\alpha/2)} + |s|_{S_T}^{(1+\alpha,\frac{1+\alpha}{2})}\right) + T^{\frac{1-\alpha}{2}}|D\underset{\sim}{u}|_{S_T}\left(|D\underset{\sim}{w}(\cdot,0)|_{\Gamma} + |s(\cdot,0)|_{\Gamma}\right).$$

Now it is not hard to prove theorem 1.2. Let us start with estimate (1.12). If we apply theorem 1.1 to problem (5.1) and take into account the boundary condition

$$s(\xi,0)|_{\Gamma} = -b(\xi,0) + 2\nu\underset{\sim}{n}_0\cdot\frac{\partial\underset{\sim}{w}_0}{\partial n}\Big|_{\Gamma},$$

we obtain

$$N \le c(T)\left(F + c_{31}\delta N + c_{32}P_T[\underset{\sim}{u}]\left(|\underset{\sim}{w}_0|_{\Omega} + (1+2\nu)|D\underset{\sim}{w}_0|_{\Omega} + |b(\cdot,0)|_{\Gamma}\right)\right)$$

where N and F are sums of norms in the left-hand side and in the right-hand side of (1.10), respectively. Assuming that

$$\delta < \frac{1}{c(T)c_{31}}$$

we obtain (1.12).

The solvability of the problem (1.7) may be proved by successive approximations. As zero approximation we take $s^{(0)} = 0$ and $\underset{\sim}{w}^{(0)}(\xi,t)$ satisfying the system $\underset{\sim}{w}_t^{(0)} - \nu\nabla^2\underset{\sim}{w}^{(0)} = \underset{\sim}{f}$, zero initial condition and the estimate

$$|\underset{\sim}{w}^{(0)}|_{Q_T}^{(2+\alpha,1+\alpha/2)} \le c_{35}|\underset{\sim}{f}|_{Q_T}^{(\alpha,\alpha/2)} \tag{5.20}$$

(for instance, we can extend $\underset{\sim}{f}$ with the preservation of class into $\mathbb{R}^3 \times$

(0,T) and solve the Cauchy problem $\underset{\sim}{w}_t^{(0)} - \nu\nabla^2\underset{\sim}{w}^{(0)} = \underset{\sim}{f},\ \underset{\sim}{w}^{(0)}\big|_{t=0} = 0$ ).

Now, we define $(\underset{\sim}{w}^{(m+1)}, s^{(m+1)})$, $m \geq 0$, as the solution of initial-boundary value problem

$$\begin{aligned}
&\underset{\sim}{w}_t^{(m+1)} - \nu\nabla^2\underset{\sim}{w}^{(m+1)} + \nabla s^{(m+1)} = \underset{\sim}{f} + \underset{\sim}{l}_1(\underset{\sim}{w}^{(m)}, s^{(m)});\\
&\nabla\cdot\underset{\sim}{w}^{(m+1)} = \rho + l_2(\underset{\sim}{w}^{(m)}),\\
&\underset{\sim}{w}^{(m+1)}\big|_{t=0} = \underset{\sim}{w}_0,\\
&\Pi_0 S(\underset{\sim}{w}^{(m+1)})\underset{\sim}{n}_0\big|_{S_T} = \Pi_0\underset{\sim}{d} + \underset{\sim}{l}_3(\underset{\sim}{w}^{(m)}),\\
&\underset{\sim}{n}_0\cdot T((\underset{\sim}{w}^{(m+1)}, s^{(m+1)})\underset{\sim}{n}_0 - \sigma\underset{\sim}{n}_0\cdot\Delta(0)\int_0^t \underset{\sim}{w}^{(m+1)}d\tau\big|_{S_T}\\
&= b + \int_0^t B\,d\tau + l_4(\underset{\sim}{w}^{(m)}, s^{(m)}) + \int_0^t l_5(\underset{\sim}{w}^{(m)})d\tau.
\end{aligned} \tag{5.21}$$

We consider this problem for $m = 0$ and verify condition iii). We have

$$\frac{\partial\rho}{\partial t} + \frac{\partial l_2(\underset{\sim}{w}^{(0)})}{\partial t} - \nabla\cdot(\underset{\sim}{f} + \underset{\sim}{l}_1(\underset{\sim}{w}^{(0)},0)) = \frac{\partial\rho}{\partial t} - \nabla_u\cdot\underset{\sim}{f} +$$

$$+\nabla\cdot(I - \mathcal{A}^*)(\underset{\sim}{w}_t^{(0)} - \underset{\sim}{f}) - \nabla\cdot\mathcal{B}_t^*\underset{\sim}{w}^{(0)} - \nabla\cdot\underset{\sim}{l}_1(\underset{\sim}{w}^{(0)},0) = \nabla\cdot(\underset{\sim}{h}^{(0)} + \underset{\sim}{h}),$$

$$\underset{\sim}{h}^{(0)} = -\nu\mathcal{B}^*\nabla^2\underset{\sim}{w}^{(0)} - \mathcal{B}_t^*\underset{\sim}{w}^{(0)} - \underset{\sim}{l}_1(\underset{\sim}{w}^{(0)},0) = \sum_{k=1}^{3}\frac{\partial\underset{\sim}{h}_k^{(0)}}{\partial\xi_k},$$

$$\underset{\sim}{h}_k^{(0)} = -\nu\mathcal{B}^*\frac{\partial\underset{\sim}{w}^{(0)}}{\partial\xi_k} - \underset{\sim}{L}_{1k}(\underset{\sim}{w}^{(0)},0) + \frac{\partial\underset{\sim}{V}^{(0)}}{\partial\xi_k},$$

$$\underset{\sim}{V}^{(0)} = -\frac{1}{4\pi}\int_\Omega\left(\nu\sum_{m=1}^{3}\frac{\partial\mathcal{B}}{\partial\xi_m}\frac{\partial\underset{\sim}{w}^{(0)}}{\partial\xi_m} - \mathcal{B}_t^*\underset{\sim}{w}^{(0)}\right)\frac{d\eta}{|\xi - \eta|}.$$

In virtue of theorem 1.1, the problem (5.21) with $m = 0$ is solvable and

$$N[\underset{\sim}{w}^{(1)}, s^{(1)}] \equiv |\underset{\sim}{w}^{(1)}|_{Q_T}^{(2+\alpha,1+\alpha/2)} + |\nabla s^{(1)}|_{Q_T}^{(\alpha,\alpha/2)}$$

$$+ |s^{(1)}|_{S_T}^{(1+\alpha,\frac{1+\alpha}{2})} + |s^{(1)}|_{Q_T}^{(1+\alpha,\gamma)} \leq C'(T)F_1$$

$$+ C(T)\left[|\underset{\sim}{l}_1(\underset{\sim}{w}^{(0)},0)|_{Q_T}^{(\alpha,\alpha/2)} + |\nabla \underset{\sim}{l}_2(\underset{\sim}{w}^{(0)})|_{Q_T}^{(\alpha,\alpha/2)}\right.$$

$$+ |\underset{\sim}{l}_2(\underset{\sim}{w}^{(0)})|_{Q_T}^{(1+\alpha,\gamma)} + |\underset{\sim}{l}_2(\underset{\sim}{w}^{(0)})|_{S_T}^{(1+\alpha,\frac{1+\alpha}{2})}$$

$$+ |\underset{\sim}{l}_3(\underset{\sim}{w}^{(0)})|_{S_T}^{(1+\alpha,\frac{1+\alpha}{2})} + |\underset{\sim}{l}_4(\underset{\sim}{w}^{(0)},0)|_{S_T}^{(1+\alpha,\frac{1+\alpha}{2})}$$

$$+ |\underset{\sim}{l}_5(\underset{\sim}{w}^{(0)})|_{S_T}^{(\alpha,\alpha/2)} + |\underset{\sim}{h}^{(0)}|_{Q_T}^{(\alpha,\alpha/2)} + \sum_{k=1}^{3}|\underset{\sim}{h}_k^{(0)}|_{Q_T}^{(1+\alpha,\gamma)}\right],$$

where

$$F_1 = F + P_T[\underset{\sim}{u}](|\underset{\sim}{w}_0|_\Omega + |D\underset{\sim}{w}_0|_\Omega + |b(\cdot,0)|_\Gamma).$$

But the sum of norms in brackets is less than

$$C_{31}\delta|\underset{\sim}{w}^{(0)}|_{Q_T}^{(2+\alpha,1+\alpha/2)} \le C_{31}C_{35}\delta|\underset{\sim}{f}|_{Q_T}^{(\alpha,\alpha/2)},$$

so

$$N[\underset{\sim}{w}^{(1)},s^{(1)}] \le C(T)(1 + C_{31}C_{35}\delta)F_1.$$

Repeating these arguments, we can show that the problem (5.21) is solvable also for $m = 1,2,\ldots$ . The differences $\underset{\sim}{z}^{(m+1)} = \underset{\sim}{w}^{(m+1)} - \underset{\sim}{w}^{(m)}$, $r^{(m+1)} = s^{(m+1)} - s^{(m)}$ (m ≥ 1) are solution to the problems

$$\underset{\sim}{z}_t^{(m+1)} - \nu\nabla^2\underset{\sim}{z}^{(m+1)} + \nabla r^{(m+1)} = \underset{\sim}{l}_1(\underset{\sim}{z}^{(m)},r^{(m)}),$$

$$\nabla\cdot\underset{\sim}{z}^{(m+1)} = l_2(\underset{\sim}{z}^{(m)})$$

$$\underset{\sim}{z}^{(m+1)}\Big|_{t=0} = 0, \qquad\qquad (5.22)$$

$$\Pi_0 S(\underset{\sim}{z}^{(m+1)})\underset{\sim}{n}_0\Big|_{S_T} = \underset{\sim}{l}_3(\underset{\sim}{z}^{(m)}),$$

$$\underset{\sim}{n}_0\cdot T(\underset{\sim}{z}^{(m+1)},r^{(m+1)})\underset{\sim}{n}_0 - \sigma\underset{\sim}{n}_0\cdot\Delta(0)\int_0^t\underset{\sim}{z}^{(m+1)}d\tau = l_4(\underset{\sim}{z}^{(m)},r^{(m)}) + \int_0^t\underset{\sim}{l}_5(\underset{\sim}{z}^{(m)})d\tau.$$

Since

$$\frac{\partial}{\partial t}l_2(\underset{\sim}{z}^{(m)}) - \nabla\cdot\underset{\sim}{l}_1(\underset{\sim}{z}^{(m)},r^{(m)}) = \nabla\cdot\underset{\sim}{h}^{(m)},$$

$$\underset{\sim}{h}^{(m)} = -\mathcal{B}^* \underset{\sim t}{z}^{(m)} - \mathcal{B}^*_t \underset{t\sim}{z}^{(m)} - \underset{\sim 1}{l}(\underset{\sim}{z}^{(m)}, r^{(m)}) = \sum_{k=1}^{3} \frac{\partial \underset{\sim k}{h}^{(m)}}{\partial \xi_k},$$

$$\underset{\sim k}{h}^{(m)} = -\nu \mathcal{B}^* \frac{\partial \underset{\sim}{z}^{(m)}}{\partial \xi_k} - \mathcal{B}^* \underset{\sim k}{e} r^{(m)} - \mathcal{B}^* \underset{\sim 1k}{L}(\underset{\sim}{z}^{(m-1)}, r^{(m-1)})$$

$$- \underset{\sim 1k}{L}(\underset{\sim}{z}^{(m)}, r^{(m)}) + \frac{\partial \underset{\sim}{V}^{(m)}}{\partial \xi_k},$$

$$\underset{\sim}{V}^{(m)} = -\frac{1}{4\pi} \int_\Omega \frac{1}{|\xi - \eta|} \left[ \sum_{j=1}^{3} \left( \nu \frac{\partial \mathcal{B}^*}{\partial \eta_j} \frac{\partial \underset{\sim}{z}^{(m)}}{\partial \eta_j} - \frac{\partial \mathcal{B}^*}{\partial \eta_j} \underset{\sim j}{e} r^{(m)} \right.\right.$$

$$\left.\left. + \frac{\partial \mathcal{B}^*}{\partial \eta_j} \underset{\sim 1j}{L}(\underset{\sim}{z}^{(m-1)}, r^{(m-1)}) \right) - \mathcal{B}^*_t \underset{t\sim}{z}^{(m)} \right] d\eta,$$

the application of theorem 1.1 to these problems yields

$$N\left[(\underset{\sim}{z}^{(m+1)}, r^{(m+1)})\right] \le C(T)\left\{ C_{31}\delta\left(N[\underset{\sim}{z}^{(m)}, r^{(m)}] + \right.\right.$$

$$\left.\left. + N[\underset{\sim}{z}^{(m-1)}, r^{(m-1)}]\right) + (\delta_{m1} + \delta_{m2}) P_T[\underset{\sim}{u}]\left( |D\underset{\sim 0}{w}|_\Omega + |\underset{\sim 0}{w}|_\Omega + |b(\cdot,0)|_\Gamma \right) \right\} \tag{5.23}$$

(we have taken into account that $\underset{\sim}{z}^{(1)} = \underset{\sim}{w}^{(1)} - \underset{\sim}{w}^{(0)}$ satisfies inhomogeneous initial condition $\underset{\sim}{z}^{(1)}\big|_{t=0} = \underset{\sim}{w}^{(0)}$).

Taking the sum of both sides in (5.23) we see that

$$\sum_M = \sum_{m=1}^{M} N[\underset{\sim}{z}^{(m+1)}, r^{(m+1)}]$$

satisfies the inequality

$$\sum_M \le 2C(T)\left\{ \sum_M \delta c_{31} + P_T[u]\left( |D\underset{\sim 0}{w}|_\Omega + |w_0|_\Omega + |b(\cdot,0)|_\Gamma \right) \right.$$

$$\left. + 2\delta N[\underset{\sim}{z}^{(1)}, s^{(1)}] + \delta N[\underset{\sim}{w}^{(0)}, 0] \right\}.$$

This shows that in the case

$$2C(T)c_{31}\delta < 1 \tag{5.24}$$

the series

$$\sum_{m=1}^{\infty} N[\underset{\sim}{z}^{(m+1)}, r^{(m+1)}]$$

is convergent which implies the convergence of $(\underset{\sim}{w}^{(m)}, s^{(m)})$. Clearly, $(\underset{\sim}{w}, s)$

$= \lim_{m\to\infty} (\underset{\sim}{w}^{(m)}, s^{(m)})$ is a solution of (1.7). The proof of theorem 1.2 is completed.

### §6. Proof of the theorem 1.3.

First of all, we prove some auxiliary estimates. In this section we shall need estimates of differences of expressions $l_1$ which are determined by two different vector fields $\underset{\sim}{u}$ and $\underset{\sim}{u}'$ sasfying conditions $(1.11_1)$, $(1.11_2)$. We denote them by $l_1$ and $l_1'$ and we set $\tilde{l}_1 = l_1 - l_1'$.

Estimates of $\tilde{l}_1$ are based on estimates of differences

$$\tilde{B}_{ij} = B_{ij} - B'_{ij}.$$

It is very easy to show that $\tilde{B}_{ij}$ satisfy $(5.9_1)$-$(5.9_4)$ with $\tilde{b}_{km} = b_{km} - b'_{km}$ instead of $b_{km}$ in right-hand sides. In their turn, $\tilde{b}_{km}$ satisfy the inequality

$$\sup_{t\le T} |\tilde{b}_{km}(\cdot,t)|^{(1+\alpha)}_{\Omega} + |\tilde{b}_{km}|^{(1+\alpha,\gamma)}_{Q_T} \le m[\underset{\sim}{u} - \underset{\sim}{u}']$$

with

$$m[\underset{\sim}{v}] = \int_0^T |D\underset{\sim}{v}|^{(1+\alpha)}_{\Omega} dt + T^{\frac{1-\alpha+\gamma}{2}} \sup_{t\le T} |D\underset{\sim}{v}|^{(\gamma)}_{\Omega}$$

that can be considered as an analogue to $(5.10_1)$ and $(5.10_2)$. Similar inequality holds for

$$\tilde{d}_{k\alpha} = d_{k\alpha} - d'_{k\alpha} = \tilde{b}_{k\alpha} + \varphi_{\alpha}\tilde{b}_{k3}\big|_{y_3=\varphi(y')}.$$

As a consequence, $\tilde{l}_1$ satisfy the inequalities (5.12) - (5.14) in which the parameter $\delta$ is replaced by $m[\underset{\sim}{u} - \underset{\sim}{u}']$ and $|D\underset{\sim}{u}|_{S_T}$ is replaced by

$$|D(\underset{\sim}{u} - \underset{\sim}{u}')|_{S_T}.$$

This is easily verified following the proof of lemma 5.2.

The derivatives $\dfrac{\partial \tilde{B}_{ij}}{\partial t}$ also satisfy $(5.9_5)$-$(5.9_7)$ with $\underset{\sim}{u}$ changed for $\underset{\sim}{u} - \underset{\sim}{u}'$, namely

$$\left|\frac{\partial \tilde{B}_{ij}}{\partial t}\right|_{Q_T}^{(\alpha,\alpha/2)} \le c_1 |D(\underset{\sim}{u} - \underset{\sim}{u}')|_{Q_T}^{(\alpha,\alpha/2)},$$

$$\langle\frac{\partial \tilde{B}_{ij}}{\partial t}\rangle_{t,Q_T}^{\left(\frac{1+\alpha-\gamma}{2}\right)} \le c_2\left(\langle D(\underset{\sim}{u} - \underset{\sim}{u}')\rangle_{t,Q_T}^{\left(\frac{1+\alpha-\gamma}{2}\right)} + |D(\underset{\sim}{u} - \underset{\sim}{u}')|_{Q_T}\right), \tag{6.1}$$

$$\left|\frac{\partial \tilde{B}_{ij}}{\partial t}\right|_{Q_T}^{(1+\alpha,\gamma)} \le c_2\left(|D(\underset{\sim}{u} - \underset{\sim}{u}')|_{Q_T}^{(1+\alpha,\gamma)} + \sup_{t\le T} |D(\underset{\sim}{u} - \underset{\sim}{u}')|_{\Omega}^{(\gamma)}\right).$$

Indeed, when we differentiate (5.8) and take the difference

$$\frac{\partial B_{ij}}{\partial t} - \frac{\partial B'_{ij}}{\partial t},$$

we see that

$$\left|\frac{\partial \tilde{B}_{ij}}{\partial t}\right|_{Q_T}^{(\alpha,\alpha/2)} \le \max_{k,m} \left|\frac{\partial \tilde{b}_{km}}{\partial t}\right|_{Q_T}^{(\alpha,\alpha/2)} (2 + 4\max |b_{km}|$$

$$+ 4\max |b'_{km}| + 4\max \left|\frac{\partial b_{km}}{\partial t}\right|_{Q_T} \max \langle\tilde{b}_{km}\rangle_{Q_T}^{(\alpha,\alpha/2)}$$

$$+ 4\max \left|\frac{\partial b'_{km}}{\partial t}\right|_{Q_T} \max \langle\tilde{b}_{km}\rangle_{Q_T}^{(\alpha,\alpha/2)}$$

$$\le |D(\underset{\sim}{u} - \underset{\sim}{u}')|_{Q_T}^{(\alpha,\alpha/2)}(2 + 8\delta) + 4|D\underset{\sim}{u}|_{Q_T} (T + T^{(1-\alpha/2)})\sup_{t\le T} |D(\underset{\sim}{u} - \underset{\sim}{u}')|_{S_t}$$

$$\le c_1 |D(\underset{\sim}{u} - \underset{\sim}{u}')|_{Q_T}^{(\alpha,\alpha/2)};$$

other estimates (5.1) are proved in the same way. The function $\underset{\sim}{l}_6(\underset{\sim}{w},s)$ and $\underset{\sim}{L}_{6m}(\underset{\sim}{w},s)$ satisfy (5.15), (5.16) with $\delta$ replaced by $m[u - u']$ and with $D\underset{\sim}{u}$ replaced by $D(\underset{\sim}{u} - \underset{\sim}{u}')$. Instead of lemma 5.4, we have

**Lemma 6.1** *If* $\underset{\sim}{u}$, $\underset{\sim}{u}'$ *satisfy* $(1.11_1)$, $(1.11_2)$ *and* $\Gamma \in C^{2+\alpha}$, *then*

$$|\tilde{I}_1(w,s)|_{Q_T}^{(\alpha,\alpha/2)} + |\nabla\tilde{I}_2(w)|_{Q_T}^{(\alpha,\alpha/2)} + |\tilde{I}_2(w)|_{Q_T}^{(1+\alpha,\gamma)}$$

$$+ \tilde{I}_6(w,s)|_{Q_T}^{(\alpha,\alpha/2)} + \sum_{m=1}^{3} |\tilde{L}_{6m}(w,s)|_{Q_T}^{(1+\alpha,\gamma)} + |\tilde{I}_2(w)|_{S_T}^{(1+\alpha,\frac{1+\alpha}{2})}$$

$$+|\tilde{I}_3(w)|_{S_T}^{(1+\alpha,\frac{1+\alpha}{2})} + |I_4(w,s)|_{S_T}^{(1+\alpha,\frac{1+\alpha}{2})} + |\tilde{I}_5(w)|_{S_T}^{(\alpha,\alpha/2)} \qquad (6.2)$$

$$\le c_3 m[u - u']\Big(|w|_{Q_T}^{(2+\alpha,1+\alpha/2)} + |\nabla s|_{Q_T}^{(\alpha,\alpha/2)} + |s|_{S_T}^{(1+\alpha,\frac{1+\alpha}{2})}$$

$$+ |s|_{Q}^{(1+\alpha,\gamma)}\Big) + c_4 P_T[u - u']\Big\{|w(\cdot,0)|_\Omega + |Dw(\cdot,0)|_\Omega + |s(\cdot,0)|_\Gamma\Big\}.$$

**Lemma 6.2** *If $u$, $u'$ satisfy $(1.11_1)$ and $(1.11_2)$, and $\Gamma \in C^{2+\alpha}$, then*

$$|n_0\cdot(\dot{\Delta}(t) - \dot{\Delta}'(t))\xi|_{S_T}^{(\alpha,\alpha/2)} \le c_5 |D(u - u')|_{S_T}^{(\alpha,\alpha/2)} \qquad (6.3)$$

*where $\Delta'(t)$ is the Laplace–Beltrami operator on $\Gamma_t' = \Omega_{u'}\Gamma$.*

Proof. Consider the function $n_0\cdot\dot{\Delta}(t)\xi$ in local coordinates $(y_1,y_2)$ in the neighbourhood of arbitrary $\xi_0 \in \Gamma$. We have

$$n_0\cdot\dot{\Delta}(t)\xi = \frac{1}{\sqrt{1+|\nabla\varphi|^2}}\left(\sum_{\gamma=1}^{2}\varphi_{y_\gamma}\dot{\Delta}(t)y_\gamma - \dot{\Delta}(t)\varphi\right)$$

$$= \frac{1}{\sqrt{1+|\nabla\varphi|^2}}\left(\sum_{\gamma=1}^{2}\varphi_{y_\gamma}\dot{h}_\gamma - \sum_{\alpha,\beta=1}^{2}\dot{g}^{\alpha\beta}\varphi_{y_\alpha y_\beta} - \sum_{\beta=1}^{2}\dot{h}_\beta\varphi_{y_\beta}\right)$$

$$= -\frac{1}{\sqrt{1+|\nabla\varphi|^2}}\sum_{\alpha,\beta=1}^{2}\dot{g}^{\alpha\beta}\varphi_{y_\alpha y_\beta}, \quad \dot{g}^{\alpha\beta} = \frac{\partial g^{\alpha\beta}}{\partial t}$$

Hence,

$$|\underset{\sim}{n}_0\cdot(\dot{\Delta}(t)-\dot{\Delta}'(t))\underset{\sim}{\xi}|^{(\alpha,\alpha/2)}_{K_d\times(0,T)} \le c_6\sum_{\gamma,\beta=1}^{2}|\dot{g}^{\gamma\beta}-\dot{g}'^{\gamma\beta}|^{(\alpha,\alpha/2)}_{K_d\times(0,T)}$$

$$\le c_7\sum_{\gamma,\beta=1}^{2}\left|\frac{\partial\omega_\gamma}{\partial y_\beta}-\frac{\partial\omega'_\gamma}{\partial y_\beta}\right|^{(\alpha,\alpha/2)}_{K_d\times(0,T)}$$

which implies (6.3).

Next, we consider

$$V(X_u,t)=\int_\Omega\frac{d\eta}{|X_u(\xi,t)-X_u(\eta,t)|}$$

and $V(X_u,t)-V(X_{u'},t)$. In Eulerian coordinates

$$V(x,t)=\int_\Omega\frac{dy}{|x-y|}.$$

It is well known (see for instance the book of N.M. Günter [G]) that

$$|V(\cdot,t)|^{(2+\alpha)}_{\Omega_t}\le c_8 \tag{6.4}$$

The constant $c_8$ may be taken independent of t, if $\underset{\sim}{u}$ satisfies $(1.11_1)$ with a small $\delta$. Since

$$c_9|\xi-\eta|\le|X_u(\xi,t)-X_u(\eta,t)|\le c_{10}|\xi-\eta|, \tag{6.5}$$

it follows that

$$|V(X_u,t)|^{(2+\alpha)}_\Omega\le c_{11}.$$

The difference $V(X_u,t)-V(X_{u'},t)\equiv\tilde{V}(\xi,t)$ can be written in the form

$$\tilde{V}(\xi,t)=\int_0^1\frac{d}{ds}V(X_{u_s},t)ds=-\int_0^1 ds\int_\Omega \underset{\sim}{H}^{(s)}(\xi,\eta,t)d\eta$$
$$\cdot\int_0^t[\underset{\sim}{\tilde{u}}(\xi,\tau)-\underset{\sim}{\tilde{u}}(\eta,\tau)]d\tau \tag{6.6}$$

where

$$\underset{\sim}{\tilde{u}}=\underset{\sim}{u}-\underset{\sim}{u}',\ \underset{\sim}{u}_s=\underset{\sim}{u}'+s\underset{\sim}{\tilde{u}},\ X_{u_s}=\underset{\sim}{\xi}+\int_0^t\underset{\sim}{u}_s(\xi,\tau)d\tau,$$

$$\underset{\sim}{H}^{(s)}(\xi,\eta,t) = \frac{X_{u_s}(\xi,t) - X_{u_s}(\eta,t)}{|X_{u_s}(\xi,t) - X_{u_s}(\eta,t)|^3}.$$

From (6.6) we obtain the formula for derivatives of $\tilde{V}$:

$$\frac{\partial\tilde{V}(\xi,t)}{\partial\xi_1} = -\int_0^t \frac{\partial\tilde{\underset{\sim}{u}}}{\partial\xi_1}d\tau\cdot\int_0^1 ds\int_\Omega \underset{\sim}{H}^{(s)}(\xi,\eta,t)d\eta$$

$$-\int_0^1 ds\int_\Omega \frac{\partial\underset{\sim}{H}^{(s)}}{\partial\xi_1}\cdot\left(\int_0^t[\tilde{\underset{\sim}{u}}(\xi,\tau) - \tilde{\underset{\sim}{u}}(\eta,\tau)]d\tau\right)d\eta.$$

We estimate $\tilde{V}$ and $\tilde{V}_{\xi_1}$ using the following elementary lemma.

**Lemma 6.3** *Let* $K(\xi,\eta)$, $\xi,\eta \in \Omega$, *satisfy the inequalities*

$$|K(\xi,\eta)| \le \frac{M}{|\xi - \eta|^2},$$

$$|K(\xi,\eta) - K(\xi',\eta)| \le M|\xi - \xi'|^\alpha\left(\frac{1}{|\xi - \eta|^{2+\alpha}} + \frac{1}{|\xi' - \eta|^{2+\alpha}}\right). \tag{6.7}$$

*Then for the function*

$$w(\xi) = \int_\Omega K(\xi,\eta)f(\eta)d\eta$$

*there holds*

$$|w|_\Omega^{(\alpha)} \le c_{12}M|f|_\Omega.$$

**Lemma 6.4** *If* $\underset{\sim}{u}$ *and* $\underset{\sim}{u}'$ *satisfy (1.11), then*

$$|\nabla\tilde{V}|_{Q_T}^{(\alpha,\alpha/2)} + \langle\tilde{V}\rangle_{t,Q_T}^{(\beta)} \le c_{13}m_1[\underset{\sim}{u} - \underset{\sim}{u}']$$

*where* $\beta \in (0,1/2]$ *and*

$$m_1[\underset{\sim}{v}] = \int_0^T |D\underset{\sim}{v}|_\Omega^{(\alpha)}dt + (T + T^{1/2})(|D\underset{\sim}{v}|_{Q_T} + |\underset{\sim}{v}|_{Q_T}).$$

Proof. We observe that in virtue of (6.5) the kernel $\underset{\sim}{H}^{(s)}(\xi,\eta,t)$

satisfies (6.7) with a constant $M > 0$ independent of $\underset{\sim}{u}$ and $\underset{\sim}{u}'$. Further, as

$$\frac{\partial \underset{\sim}{H}^{(s)}}{\partial \xi_1} = \sum_{k=1}^{3} \underset{\sim k}{H}^{(s)}(\xi,\eta,t)(\delta_{kl} + \int_0^t \frac{\partial}{\partial \xi_1} u_{sk}(\xi,\tau)d\tau),$$

with

$$\underset{\sim k}{H}^{(s)} = \frac{\underset{\sim k}{e}}{|X_{u_s}(\xi,t) - X_{u_s}(\eta,t)|^3} - 3\,\frac{X_{u_s}(\xi,t) - X_{u_s}(\eta,t)}{|X_{u_s}(\xi,t) - X_{u_s}(\eta,t)|^5}$$

$$\times (\underset{\sim k}{e}\cdot X_{u_s}(\xi,t) - X_{u_s}(\eta,t)),$$

the kernel

$$\mathcal{H}_1^{(s)} = \frac{\partial \underset{\sim}{H}^{(s)}}{\partial \xi_1}\cdot\int_0^t [\underset{\sim}{\tilde{u}}(\xi,\tau) - \underset{\sim}{\tilde{u}}(\eta,\tau)]d\tau$$

also satisfies (6.7) with a constant proportional to

$$\int_0^T (|D\underset{\sim}{\tilde{u}}|_\Omega + |\underset{\sim}{\tilde{u}}|_\Omega)d\tau.$$

Hence,

$$\sup_{t\le T} \left|\frac{\partial \tilde{V}}{\partial \xi_1}\right|_\Omega^{(\alpha)} \le c_{14} m_1[\underset{\sim}{\tilde{u}}].$$

Now, we consider the difference

$$\frac{\partial \tilde{V}(\xi,t)}{\partial \xi_1} - \frac{\partial \tilde{V}(\xi,t-\tau)}{\partial \xi_1} = -\int_{t-\tau}^{t} \frac{\partial \underset{\sim}{\tilde{u}}(\xi,\lambda)}{\partial \xi_1}\, d\lambda\cdot\int_0^1 ds\int_\Omega \underset{\sim}{H}^{(s)}(\xi,\eta,t)d\eta$$

$$-\int_0^{t-\tau} \frac{\partial \underset{\sim}{\tilde{u}}(\xi,\lambda)}{\partial \xi_1}\, d\lambda\cdot\int_0^1 ds\int_\Omega \Big[H^{(s)}(\xi,\eta,t) - H^{(s)}(\xi,\eta,t-\tau)d\eta$$

$$-\int_0^1 ds\int_\Omega \Big[\mathcal{H}_1^{(s)}(\xi,\eta,t) - \mathcal{H}_1^{(s)}(\xi,\eta,t-\tau)\Big]d\eta.$$

Elementary calculations show that the kernels

$$\underset{\sim}{H}^{(s)}(\xi,\eta,t) - H^{(s)}(\xi,\eta,t-\tau) \text{ and } \mathcal{H}_1^{(s)}(\xi,\eta,t) - \mathcal{H}_1^{(s)}(\xi,\eta,t-\tau)$$

satisfy (6.7) with constants M proportional to

$$\int_{t-\tau}^{t} (|Du_s|_\Omega + |u_s|_\Omega) d\lambda \le \tau^{\alpha/2} m_1[u_s] \le \delta\tau^{\alpha/2}$$

and to $\tau^{\alpha/2} m_1[\tilde{u}]$, respectively, so

$$\langle \frac{\partial \tilde{V}}{\partial \xi_1} \rangle^{(\alpha/2)}_{t,Q_T} \le c_{16} m_1[\tilde{u}].$$

By the same arguments it can be deduced from (6.6) that

$$\langle \tilde{V} \rangle^{(\beta)}_{t,Q_T} \le c_{16} m_1[\tilde{u}]$$

for $\beta \in (0,1/2)$. The lemma is proved.

**Corollary** *Under the condition (1.11)*

$$|\nabla V(X_u,t)|^{(\alpha,\alpha/2)}_{Q_T} \le c_{13}\delta + c_8 ,$$

$$\langle V \rangle^{(\beta)}_{t,Q_T} \le c_{13}\delta, \quad \beta \in (0,1/2].$$

Indeed, if we take $u' = 0$, then

$$V'(X_u,t) = \int_\Omega \frac{d\eta}{|\xi - \eta|} ,$$

and we get

$$|\nabla V|^{(\alpha,\alpha/2)}_{Q_T} \le |\nabla(V - V')|^{(\alpha,\alpha/2)}_{Q_T} + |\nabla V'|^{(\alpha)}_\Omega \le c_{13} m_1[u] + c_8 \le c_{13}\delta + c_8,$$

$$\langle V \rangle^{(\beta)}_{t,Q_T} = \langle V - V' \rangle^{(\beta)}_{t,Q_T} \le c_{13}\delta.$$

Finally, we estimate the difference

$$f(X_u,t) - f(X_{u'},t) = \sum_{k=1}^{3} \int_0^t (u_k - u_k')d\tau \int_0^1 f_{x_k}(X_{u_s},t)ds.$$

The following lemma is an immediate consequence of this representation formula.

**Lemma 6.5** *If $f(x,t)$ satisfies the hypotheses of theorem 1.3 and $u$, $u'$*

*satisfy (1.11), then*

$$|\underset{\sim}{f}(X_u,t) - \underset{\sim}{f}(X_{u'},t)|^{(\alpha,\alpha/2)}_{Q_T} + \langle \underset{\sim}{f}(X_u,t) - f(X_{u'},t)\rangle^{(\beta)}_{t,Q_T} \le c_{17} m_1 [\underset{\sim}{u} - \underset{\sim}{u}']$$

*where* $\beta \in (0,1/2]$.

Let us turn to the proof of theorem 1.3. First of all, we rewrite the boundary condition in (1.6). We take the projection of $\mathbb{T}_u\underset{\sim}{n} - \sigma\Delta(t)X_u$ onto the normal $\underset{\sim}{n}_0$ to $\Gamma$ and onto the tangent plane to $\Gamma_t$ at the point $X_u(\xi,t)$, and then we again project the tangential component onto the tangent plane to $\Gamma$ at $\xi$. This will give us the boundary conditions

$$\nu\Pi_0\Pi\mathbb{S}_u(\underset{\sim}{u})\underset{\sim}{n} = 0,$$

$$\underset{\sim}{n}_0\cdot\mathbb{T}_u(\underset{\sim}{u},q)\underset{\sim}{n} - \sigma\underset{\sim}{n}_0\cdot\Delta(t)X_u = 0.$$

It is easy to see that these conditions are equivalent to

$$\mathbb{T}_u\underset{\sim}{n} - \sigma\Delta(t)X_u = 0,$$

provided that

$$\underset{\sim}{n}_0\cdot\underset{\sim}{n} = |\mathcal{A}\underset{\sim}{n}_0|^{-1}(\underset{\sim}{n}_0\cdot\mathcal{A}\underset{\sim}{n}_0) > 0$$

which is certainly true if $\underset{\sim}{u}$ satisfies the condition $(1.11_1)$ with a small $\delta$. Now, since

$$\Delta(t)X_u = \Delta(t)\xi + \Delta(t)\int_0^t \underset{\sim}{u}d\tau = \Delta(0)\xi + \int_0^t \dot{\Delta}(\tau)\xi d\tau + \Delta(t)\int_0^t \underset{\sim}{u}d\tau,$$

we can write the normal component of our boundary condition in the form

$$\underset{\sim}{n}_0\cdot\mathbb{T}_u(\underset{\sim}{u},q)\underset{\sim}{n} - \sigma\underset{\sim}{n}_0\cdot\Delta(t)\int_0^t \underset{\sim}{u}d\tau = \sigma\underset{\sim}{n}_0\cdot\int_0^t \dot{\Delta}(\tau)\xi d\tau + \sigma H_0(\xi)$$

where $H_0(\xi) = \underset{\sim}{n}_0\cdot\Delta(0)\xi$ is the twice mean curvature of $\Gamma$. Thus, we have to solve the problem

$$\begin{gathered}
\underset{\sim}{u}_t - \nu\nabla_u^2\underset{\sim}{u} + \nabla_u q = \kappa\nabla_u V(X_u,t) + \underset{\sim}{f}(X_u,t), \\
\nabla_u\cdot\underset{\sim}{u} = 0 \qquad (\xi\in\Omega,\ t>0), \\
\underset{\sim}{u}(\xi,0) = \underset{\sim}{v}_0(\xi), \\
\nu\Pi_0\Pi\mathbb{S}_u(\underset{\sim}{u})\underset{\sim}{n} = 0, \\
\underset{\sim}{n}_0\cdot\mathbb{T}_u(\underset{\sim}{u},q)\underset{\sim}{n} - \sigma\underset{\sim}{n}_0\cdot\Delta(t)\int_0^t \underset{\sim}{u}d\tau = \sigma\underset{\sim}{n}_0\cdot\int_0^t \dot{\Delta}(\tau)\xi d\tau + \sigma H_0(\xi) \quad (\xi\in\Gamma).
\end{gathered} \tag{6.8}$$

We again use the method of successive approximations. We take $\underset{\sim}{u}^{(0)}=0$,

$q^{(0)} = 0$ and we define approximations $(\underset{\sim}{u}^{(m+1)}, q^{(m+1)})$ as solutions of linear problems

$$
\begin{aligned}
&\underset{\sim t}{u}^{(m+1)} - \nu\nabla^2_m \underset{\sim}{u}^{(m+1)} + \nabla_m q^{(m+1)} = \kappa\nabla_m V(X_m, t) + \underset{\sim}{f}(X_m, t),\\
&\nabla_m \cdot \underset{\sim}{u}^{(m+1)} = 0,\\
&\underset{\sim}{u}^{(m+1)}(\xi, 0) = \underset{\sim}{v}_0(\xi),\\
&\nu\Pi_0\Pi_m S_m(\underset{\sim}{u}^{(m+1)})\underset{\sim}{n}_m = 0,\\
&\underset{\sim}{n}_0 \cdot T_m(\underset{\sim}{u}^{(m+1)}, q^{(m+1)})\underset{\sim}{n}_m - \sigma \underset{\sim}{n}_0 \cdot \Delta_m(t)\int_0^t u^{(m+1)} d\tau = \sigma H_0(\xi) + \sigma\int_0^t \underset{\sim}{n}_0 \cdot \dot{\Delta}_m(\tau)\xi d\tau .
\end{aligned}
\tag{6.9}
$$

Here $X_m = X_{u^{(m)}}$, $\nabla_m = \nabla_{u^{(m)}}$, $\underset{\sim}{n}_m$ is a normal to the surface

$$\Gamma_m(t) = X_m(\Gamma),$$

$\Delta_m(t)$ is the Laplace-Beltrami operator on this surface,

$$\Pi_m \underset{\sim}{f} = \underset{\sim}{f} - \underset{\sim}{n}_m(\underset{\sim}{n}_m \cdot \underset{\sim}{f}),\ S_m = S_{u^{(m)}},\ T_m = T_{u^{(m)}}.$$

In particular, for $m = 0$ we have

$$
\begin{aligned}
&\underset{\sim t}{u}^{(1)} - \nu\nabla^2 \underset{\sim}{u}^{(1)} + \nabla q^{(1)} = \kappa\nabla V(\xi) + \underset{\sim}{f}(\xi, t),\\
&\nabla \cdot \underset{\sim}{u}^{(1)} = 0, \qquad (\xi \in \Omega,\ t > 0),\\
&\underset{\sim}{u}^{(1)}(\xi, 0) = \underset{\sim}{v}_0(\xi),\\
&\nu\Pi_0 S(\underset{\sim}{u}^{(1)})\underset{\sim}{n}_0 = 0, \qquad (\xi \in \Gamma),\\
&\underset{\sim}{n}_0 \cdot T(\underset{\sim}{u}^{(1)}, q^{(1)})\underset{\sim}{n}_0 - \sigma\underset{\sim}{n}_0 \cdot \Delta(0)\int_0^t \underset{\sim}{u}^{(1)} d\tau = \sigma H_0(\xi).
\end{aligned}
\tag{6.10}
$$

All the hypotheses of theorem 1.1 are satisfied for this problem, in particular, iii) holds with $\underset{\sim}{h} = -\underset{\sim}{f} - \kappa\nabla V$ and with

$$\underset{\sim k}{h} = -\kappa \underset{\sim k}{e} V + \frac{1}{4\pi}\frac{\partial}{\partial \xi_k}\int_{\Omega} \underset{\sim}{f}(\eta, t)\frac{d\eta}{|\xi - \eta|}$$

satisfying the inequalities

$$|\underset{\sim}{h}|^{(\alpha,\alpha/2)}_{Q_T} \le |\underset{\sim}{f}|^{(\alpha,\alpha/2)}_{Q_T} + \kappa|\nabla V|^{(\alpha)}_{\Omega},$$

$$|\underset{\sim}{h}_k|_{Q_T}^{(1+\alpha,\gamma)} \le c_{18}\langle \underset{\sim}{f}\rangle_{t,Q_T}^{\frac{1+\alpha-\gamma}{2}}$$

with $\gamma = 1 - \varepsilon$. Therefore, by theorem 1.1, the problem (6.10) is uniquely solvable for $t \in (0,T)$ and

$$N_T[\underset{\sim}{u}^{(1)},q^{(1)}] \equiv N[\underset{\sim}{u}^{(1)},q^{(1)}] \le c_{19}(T)\Big(|\underset{\sim}{f}|_{C^{\alpha,\frac{\alpha+\varepsilon}{2}}(Q_T)} + \kappa|\nabla V|_{\Omega}^{(\alpha)} + |\underset{\sim}{v}_0|^{(2+\alpha)} + |H_0|_{\Gamma}^{(1+\alpha)}\Big). \tag{6.11}$$

We can apply theorem 1.2 to show successively that all problems (6.9) are solvable on time intervals $(0,T_{m+1})$.

The interval $(0,T_{m+1})$ should possess the property that the preceding approximation $\underset{\sim}{u}^{(m)}$ is defined there and that it satisfies conditions (1.11) for $T = T_{m+1}$ with a small $\delta$, as required in theorem 1.2. We want to prove that $T_m \ge T' > 0$ for all m, that the norms $N_{T'}[\underset{\sim}{u}^{(m)},q^{(m)}]$ are uniformly bounded and that the sequence $\{\underset{\sim}{u}^{(m)},q^{(m)}\}$ is convergent to a solution of the problem (6.8).

Suppose that $\underset{\sim}{u}^{(j)},q^{(j)}$, $j = 1,\dots,m+1$ are defined for $t \in (0,T_{m+1})$. When we subtract from each other the equations (6.9) for two neighbouring indices j and j − 1, we show that the differences

$$\underset{\sim}{w}^{(j+1)} = \underset{\sim}{u}^{(j+1)} - \underset{\sim}{u}^{(j)}, \quad s^{(j+1)} = q^{(j+1)} - q^{(j)}$$

solve the initial-boundary value problem

$$\begin{aligned}
&\underset{\sim}{w}_t^{(j+1)} - \nu\nabla_j^2\underset{\sim}{w}^{(j+1)} + \nabla_j s^{(j+1)} = \underset{\sim}{l}_1^{(j)}(\underset{\sim}{u}^{(j)},q^{(j)}) - \underset{\sim}{l}_1^{(j-1)}(\underset{\sim}{u}^{(j)},q^{(j)}) + \kappa\nabla_j[V(X_j,t)\\
&\quad - V(X_{j-1},t)] + \kappa(\nabla_j - \nabla_{j-1})V(X_{j-1},t) + \underset{\sim}{f}(X_j,t) - \underset{\sim}{f}(X_{j-1},t) \equiv \underset{\sim}{g}_j(\xi,t),\\
&\nabla_j\cdot\underset{\sim}{w}^{(j+1)} = l_2^{(j)}(\underset{\sim}{u}^{(j)}) - l_2^{(j-1)}(\underset{\sim}{u}^{(j)}) \equiv r_j(\xi,t)\\
&\underset{\sim}{w}^{(j+1)}(\xi,0) = 0,\\
&\nu\Pi_0\Pi_j S_j(\underset{\sim}{w}^{(j+1)})\underset{\sim}{n}_j\big|_{\xi\in\Gamma} = \underset{\sim}{l}_3^{(j)}(\underset{\sim}{u}^{(j)}) - \underset{\sim}{l}_3^{(j-1)}(\underset{\sim}{u}^{(j)}),\\
&\underset{\sim}{n}_0\cdot T_j(\underset{\sim}{w}^{(j+1)},s^{(j+1)})\underset{\sim}{n}_j - \sigma\underset{\sim}{n}_0\cdot\Delta_j\int_0^t\underset{\sim}{w}^{(j+1)}d\tau\big|_{\xi\in\Gamma}\\
&= l_4^{(j)}(\underset{\sim}{u}^{(j)},q^{(j)}) - l_4^{(j-1)}(\underset{\sim}{u}^{(j)},q^{(j)}) + \int_0^t[l_5^{(j)}(u^{(j)})\\
&\quad - l_5^{(j-1)}(\underset{\sim}{u}^{(j)})]d\tau + \sigma\int_0^t\underset{\sim}{n}_0\cdot(\dot{\Delta}_j - \dot{\Delta}_j)\underset{\sim}{\xi}d\tau\big|_{\xi\in\Gamma}
\end{aligned} \tag{6.12}$$

where $l_i^{(j)}$ are expressions $l_i$ (5.2) corresponding to the vector fields $\underset{\sim}{u} = \underset{\sim}{u}^{(j)}$.

Let us write the representation formulas iii). We have

$$\frac{\partial}{\partial t} r_j(\xi,t) - \nabla_j g_j = \nabla \cdot \underset{\sim}{H}_j,$$

$$\underset{\sim}{H}_j = \frac{\partial}{\partial t}\left(\mathcal{B}^*_{j-1} - \mathcal{B}^*_j\right)\underset{\sim}{u}^{(j)} - \mathcal{A}^*_j g_j = \left(\mathcal{B}^*_{j-1} - \mathcal{B}^*_j\right)\underset{\sim}{u}_t^{(j)}$$

$$+ \left(\mathcal{B}^*_{j-1} - \mathcal{B}^*_j\right)_t \underset{\sim}{u}^{(j)} - \mathcal{A}^*_j g_j.$$

As the vector field $g_j$ can be written in the form

$$g_j = \sum_{k=1}^{3} \frac{\partial}{\partial \xi_k} \underset{\sim}{G}_{jk} + [\underset{\sim}{f}(X_j,t) - \underset{\sim}{f}(X_{j-1},t)]$$

with

$$\underset{\sim}{G}_{jk} = \underset{\sim}{L}_{1k}^{(j)}(\underset{\sim}{u}^{(j)},q^{(j)}) - \underset{\sim}{L}_{1k}^{(j-1)}(\underset{\sim}{u}^{(j)},q^{(j)}) + \kappa\mathcal{A}_j \underset{\sim}{e}_k[V(X_j,t)$$

$$- V(X_{j-1},t)] + \kappa(\mathcal{A}_j - \mathcal{A}_{j-1})\underset{\sim}{e}_k V(X_{j-1},t)$$

and

$$\underset{\sim}{u}_t^{(j)} = \nu\nabla^2_{j-1}\underset{\sim}{u}^{(j)} - \nabla_{j-1}q^{(j)} + \kappa\nabla_{j-1}V(X_{j-1},t) + \underset{\sim}{f}(X_{j-1},t)$$

$$= \sum_{k=1}^{3} \frac{\partial}{\partial \xi_k} \underset{\sim}{M}_{jk}(\xi,t) + \underset{\sim}{f}(X_{j-1},t)$$

with

$$\underset{\sim}{M}_{jk}(\xi,t) = \nu(\mathcal{A}_{j-1}\underset{\sim}{e}_k \cdot \nabla_{j-1})\underset{\sim}{u}^{(j)} - \mathcal{A}_{j-1}e_k q^{(j)} + \kappa\mathcal{A}_{j-1}\underset{\sim}{e}_k V(X_{j-1},t)$$

it follows that

$$\underset{\sim}{H}_j = \sum_{k=1}^{3} \frac{\partial \underset{\sim}{H}_{jk}}{\partial \xi_k},$$

$$\underset{\sim}{H}_{jk} = (\mathcal{B}^*_{j-1} - \mathcal{B}^*_j)\underset{\sim}{M}_{jk}(\xi,t) - \mathcal{A}^*_j G_{jk}(\xi,t) + \frac{\partial \underset{\sim}{W}_j}{\partial \xi_k},$$

where $\underset{\sim}{W}_j$ is a potential

$$W_j(\xi,t) = +\frac{1}{4\pi}\int_\Omega\left\{\sum_{l=1}^{3}\left(\frac{\partial(\mathcal{B}^*_{j-1} - \mathcal{B}^*_j)}{\partial\eta_l} M_{jl}(\eta,t) - \frac{\partial\mathcal{A}^*_j}{\partial\eta_l} G_{jl}\right)\right.$$

$$\left. - (\mathcal{B}^*_j - \mathcal{B}^*_{j-1}) f(X_j,t) - (\mathcal{B}^*_{j-1} - \mathcal{B}^*_j) u^{(j)}_t + \mathcal{A}^*_j (f(X_j,t) - f(X_{j-1},t))\right\} \frac{d\eta}{|\xi - \eta|}.$$

We now proceed to the estimates of $w^{(j+1)}$ and $s^{(j+1)}$. We suppose that $u^{(1)},\ldots,u^{(m)}$ satisfy the condition $(1.11_1)$ for $T \le T_{m+1}$; moreover, we assume a stronger condition

$$(T + T^{1/2})N_T[u^{(j)},q^{(j)}] \le \delta \quad (T \le T_{(m+1)}, \quad j = 1,\ldots,m). \tag{6.13}$$

The inequality $(1.11_2)$ is a consequence of $(1.11_1)$ since $\gamma = 1 - \varepsilon$ and

$$\frac{1 - \alpha + \gamma}{2} = 1 - \frac{\alpha + \varepsilon}{2} \ge \frac{1}{2}.$$

We apply lemma 6.4 and lemma 6.5 to estimate the terms in $g_j$ containing $V$ and $f$ and we estimate the difference $l^{(j)} - l^{(j-1)}$ by inequality (6.2). As for $H$ and $H_{jk}$, they have the same structure as

$$l_6^{(j)}(u^{(j)},s^{(j)}) - l_6^{(j-1)}(u^{(j)},s^{(j)})$$

and

$$L_{6k}^{(j)}(u^{(j)},s^{(j)}) - L_{6k}^{(j-1)}(u^{(j)},s^{(j)})$$

and they also can be estimated by inequality (6.2). This gives

$$|g_j|_{Q_T}^{(\alpha,\alpha/2)} + |\nabla r_j|_{Q_T}^{(\alpha,\alpha/2)} + |r_j|_{Q_T}^{(1+\alpha,\gamma)} + |r_j|_{S_T}^{(1+\alpha,\frac{1+\alpha}{2})}$$
$$+ \left|l_3^{(j)}(u^{(j)}) - l_3^{(j-1)}(u^{(j)})\right|_S^{(1+\alpha,\frac{1+\alpha}{2})} + \left|l_4^{(j)}(u^{(j)},q^{(j)})\right.$$
$$\left. - l_4^{(j-1)}(u^{(j)},q^{(j)})\right|_{S_T}^{(1+\alpha,\frac{1+\alpha}{2})} + |l_5^{(j)}(u)^{(j)} - l_5^{(j-1)}(u^{(1j)})|_{S_T}^{(\alpha,\alpha/2)}$$
$$+ |H_j|_{Q_T}^{(\alpha,\alpha/2)} + \sum_{k=1}^{3}|H_{jk}|_{Q_T}^{(1+\alpha,\gamma)} \le c_{20}\Big\{m[u^{(j)} - u^{(j-1)}]N_T[u^{(j)},q^{(j)}]$$
$$+ m_1[u^{(j)} - u^{(j-1)}]\Big\} + c_5|D(u^{(j)} - u^{(j-1)})|_{Q_T}^{(\alpha,\alpha/2)}$$
$$+ c_{21}P_T[u^{(j)} - u^{(j-1)}](|u^{(j)}(\cdot,0)|_\Omega + |Du^{(j)}(\cdot,0)|_\Omega + |q^{(j)}(\cdot,0)|_\Gamma). \tag{6.14}$$

By the assumption (6.13),

$$N_T[\underset{\sim}{u}^{(j)},q^{(j)}]m[\underset{\sim}{u}^{(j)} - \underset{\sim}{u}^{(j-1)}]$$

$$\le 2(T + T^{1/2})|\underset{\sim}{u}^{(j)} - \underset{\sim}{u}^{(j-1)}|_{Q_T}^{(2+\alpha,1+\alpha/2)} N_T[\underset{\sim}{u}^{(j)},q^{(j)}]$$

$$\le 2\delta|\underset{\sim}{u}^{(j)} - \underset{\sim}{u}^{(j-1)}|_{Q_T}^{(2+\alpha,1+\alpha/2)}$$

Moreover, if $j \ge 2$, then $\underset{\sim}{u}^{(j)} - \underset{\sim}{u}^{(j-1)}\Big|_{t=0} = 0$, and

$$P_T[\underset{\sim}{u}^{(j)} - \underset{\sim}{u}^{(j-1)}] \le c_{22}(T + T^{1/2})|\underset{\sim}{u}^{(j)} - \underset{\sim}{u}^{(j-1)}|_{Q_T}^{(2+\alpha,1+\alpha/2)}$$

so the last term in (6.14) does not exceed

$$c_{21}c_{22}\delta|\underset{\sim}{u}^{(j)} - \underset{\sim}{u}^{(j-1)}|_{Q_T}^{(2+\alpha,1+\alpha/2)}$$

For $j = 1$ we have $\underset{\sim}{u}^{(j)} - \underset{\sim}{u}^{(j-1)} = \underset{\sim}{u}^{(1)}$,

$$P_T[\underset{\sim}{u}^{(1)}] \le (1 + T^{\frac{1-\alpha}{2}})|\underset{\sim}{u}^{(1)}(\cdot,0)|_\Omega^{(\alpha+1)} + c_{22}(T + T^{1/2})|\underset{\sim}{u}^{(1)}|_{Q_T}^{(2+\alpha,1+\alpha/2)}$$

and last term in (6.14) is less than

$$c_{21}c_{22}\delta|\underset{\sim}{u}^{(1)}|_{Q_T}^{(2+\alpha,1+\alpha/2)} + c_{23}(1 + T^{\frac{1-\alpha}{2}})(|\underset{\sim}{v}_0|_\Omega^{(1+\alpha)}|\underset{\sim}{v}_0|_\Omega$$

$$+ |D\underset{\sim}{v}_0|_\Omega + |\sigma|_\Gamma|H_0|_\Gamma),$$

since

$$\underset{\sim}{u}^{(1)}\Big|_{t=0} = \underset{\sim}{v}_0 \quad\text{and}\quad q^{(1)}\Big|_{t=0} = 2\nu\underset{\sim}{n}_0\cdot\frac{\partial\underset{\sim}{v}_0}{\partial n}\Big|_{\xi\in\Gamma} - \sigma H_0(\xi).$$

So, in virtue of inequality (1.12) applied to the problem (6.12), we have

$$N_T[\underset{\sim}{w}^{(1+j)},s^{(1+j)}] \le c(T)\Big\{(2c_{20} + c_{21}c_{22})\delta N_T[\underset{\sim}{w}^{(j)},s^{(j)}]$$

$$+ \delta_{j1}c_{23}(1 + T^{\frac{1-\alpha}{2}})|\underset{\sim}{v}_0|_\Omega^{(1+\alpha)}(|\underset{\sim}{v}_0|_\Omega + |D\underset{\sim}{v}_0|_\Omega + |\sigma|_\Gamma|H_0|_\Gamma)$$

$$+ c_{20}m_1[\underset{\sim}{w}^{(j)}] + c_5|D\underset{\sim}{w}^{(j)}|_{Q_T}^{(\alpha,\alpha/2)}\Big\}, \quad T \le T_{m+1}.$$

Let us estimate the norms of $\tilde{w}^{(j)}$ on the right-hand side. First of all, we observe that

$$m_1[\tilde{w}^{(j)}] \le (T + T^{1/2})\Big(|D\tilde{w}^{(j)}|^{(\alpha,\alpha/2)}_{Q_T} + |\tilde{w}^{(j)}|_{Q_T}\Big),$$

and the norm $|D\tilde{w}^{(j)}|^{(\alpha,\alpha/2)}_{Q_T}$ can be estimated by the following interpolation inequalities;

$$\langle D\tilde{w}^{(j)}\rangle^{(\alpha/2)}_{t,Q_T} \le \left(\langle D\tilde{w}^{(j)}\rangle^{\left(\frac{1+\alpha}{2}\right)}_{t,Q_T}\right)^{\frac{\alpha}{1+\alpha}}\left(2|D\tilde{w}^{(j)}|_{Q_T}\right)^{\frac{1}{1+\alpha}}$$

$$\le \varepsilon_1|\tilde{w}^{(j)}|^{(2+\alpha,1+\alpha/2)}_{Q_T} + \varepsilon_1^{-\alpha}\, 2|D\tilde{w}^{(j)}|_{Q_T},$$

$$|D\tilde{w}^{(j)}|^{(\alpha)}_{\Omega} \le \varepsilon_2|\tilde{w}^{(j)}|^{(2+\alpha)}_{\Omega} + c_{24}\varepsilon_2^{-1-\alpha}|\tilde{w}^{(j)}|_{\Omega}.$$

With the appropriate choice of $\varepsilon_1$ and $\varepsilon_2$, we obtain

$$N_T[\tilde{w}^{(1+j)},s^{(1+j)}] \le \Big[c(T)(2c_{20} + c_{21}c_{22})\delta + \varepsilon\Big]N_T[\tilde{w}^{(j)},s^{(j)}]$$

$$+ c_{25}(\varepsilon)|\tilde{w}^{(j)}|_{Q_T} + \delta_{j1}c(T)c_{23}(1 + T^{\frac{1-\alpha}{2}})|\tilde{v}_0|^{(1+\alpha)}_{\Omega}(|\tilde{v}_0|_{\Omega} + |D\tilde{v}_0|_{\Omega} \tag{6.15}$$

$$+ |\sigma|_\Gamma|H_0|_\Gamma), \quad \varepsilon \in (0,1), \quad T \le T_{m+1}$$

Suppose that $\delta$ is small enough and choose $\varepsilon$ small to satisfy the inequality

$$\mu \equiv c(T)(2c_{20} + c_{21}c_{22})\delta + \varepsilon < 1$$

Let

$$\sum_{m+1}(T) = \sum_{j=1}^{m+1} N_T[\tilde{w}^{(j)},s^{(j)}].$$

Since

$$|\tilde{w}^{(j)}|_{Q_T} \le |\tilde{w}^{(j)}(\cdot,0)|_{\Omega} + \int_0^T|\tilde{w}_t^{(j)}(\cdot,t)dt \le |\tilde{v}_0|_{\Omega} + \int_0^T N_t[\tilde{w}^{(j)},s^{(j)}]dt,$$

the summation of (6.15) with respect to $j = 1,\ldots,m$ gives

$$\sum_{m+1}(T) \le \mu\sum_{m+1}(T) + c_{25}\int_0^T\sum_{m+1}(t)dt + \mathcal{F}_2(T)$$

where

$$\mathcal{F}_2(T) = N_T[\underset{\sim}{w}^{(1)},s^{(1)}] + c_{25}|\underset{\sim 0}{v}|_\Omega$$

$$+ c(T)c_{23}(1 + T^{\frac{1-\alpha}{2}})|\underset{\sim 0}{v}|_\Omega^{(1+\alpha)}(|\underset{\sim 0}{v}|_\Omega + |D\underset{\sim 0}{v}|_\Omega + |\sigma|_\Gamma|H_0|_\Gamma).$$

Finally, by the Gronwall lemma, we obtain a uniform bound

$$N_T[\underset{\sim}{u}^{(m+1)},q^{(m+1)}] \le \sum_{m+1}(T) \le \frac{c_{25}}{1-\mu}\int_0^T \mathcal{F}_2(t)e^{\frac{c_{25}}{1-\mu}(T-t)}dt$$

$$+ \frac{\mathcal{F}_2(T)}{1-\mu} \le \frac{\mathcal{F}_2(T)}{1-\mu}e^{\frac{c_{25}T}{1-\mu}}.$$

The condition (6.13) for $\underset{\sim}{u}^{(m+1)},q^{(m+1)}$ will be satisfied, if

$$(T + T^{1/2})\frac{\mathcal{F}_2(T)}{1-\mu}e^{\frac{c_{25}T}{1-\mu}} \le \delta \tag{6.17}$$

The left-hand side of this inequality does not depend on m, and we can conclude that there exists a time interval $(0,T')$ where all the approximations are defined and satisfy (6.13). The inequality (6.16) shows that the series

$$\sum_{j=1}^{\infty} N_T[w^{(j)}s^{(j)}]$$

is convergent, hence the sequence $\{\underset{\sim}{u}^{(m)},q^{(m)}\}$ is convergent in the norm $N_{T'}$. Passage to the limit in (6.9) proves that $(\underset{\sim}{u},q) = \lim_{m\to\infty}(\underset{\sim}{u}^{(m)},q^{(m)})$ is a solution of the problem (6.8).

To prove the uniqueness we observe that the difference of the two solutions $\underset{\sim}{w} = \underset{\sim}{u} - \underset{\sim}{u}'$, $s = q - q'$ solves the initial boundary value problem of the type (6.12), namely,

$$\underset{\sim t}{w} - \nu\nabla_u^2\underset{\sim}{w} + \nabla_u s = \underset{\sim 1}{l}(\underset{\sim}{u}',q') - \underset{\sim 1}{l}'(\underset{\sim}{u}',q') + \kappa\nabla_u[V(X_u,t)$$

$$- V(X_{u'},t)] + \kappa(\nabla_u - \nabla_{u'})V(X_u,t) + \underset{\sim}{f}(X_u,t) - \underset{\sim}{f}(X_{u'},t),$$

$$\nabla_u \cdot \underset{\sim}{w} = l_2(\underset{\sim}{u}') - l_2'(\underset{\sim}{u}'), \quad \underset{\sim}{w}\big|_{t=0} = 0,$$

$$\nu\Pi_0\Pi S_u(\underset{\sim}{w})\underset{\sim}{n}\big|_{\xi\in\Gamma} = \underset{\sim}{l}_3(\underset{\sim}{u}') - \underset{\sim}{l}_3'(\underset{\sim}{u}'),$$

$$\underset{\sim}{n}_0\cdot T_u(\underset{\sim}{w},s)\underset{\sim}{n} - \sigma\underset{\sim}{n}_0\cdot\Delta(t)\int_0^t \underset{\sim}{w}d\tau\big|_{\xi\in\Gamma} = l_4(\underset{\sim}{u}',q') - l_4'(\underset{\sim}{u}',q')$$

$$+ \int_0^t (l_5(\underset{\sim}{u}') - l_5'(\underset{\sim}{u}'))d\tau + \sigma\int_0^t \underset{\sim}{n}_0\cdot(\mathring{\Delta}(\tau) - \mathring{\Delta}'(\tau))\underset{\sim}{\xi}d\tau.$$

Repeating exactly the above arguments we arrive at the inequality of the type (6.15), i.e.

$$N_T[\underset{\sim}{w},s] \le \mu N_T[\underset{\sim}{w},s] + c_{25}|w|_{Q_T} \le \mu N_T[\underset{\sim}{w},s] + c_{25}\int_0^T N_t[\underset{\sim}{w},s]dt$$

which implies $\underset{\sim}{w} = 0$, $s = 0$. Theorem 1.3 is proved.

## References


[ADN] Agmon S., Douglis A., Nirenberg L., *Estimates near the boundary for the solutions of elliptic partial differential equations satisfying general boundary conditions* I, Comm. Pure Appl. Math., Vol 12, 1959, p.p. 623 - 727.

[A] Allain G., *Small-time existence for the Navier-Stokes equations with a free surface*, Ecole Polytechnique, Rapport interne No 135, 1985, p.p 1 -24.

[B1] Beale J. Th., *The initial-value problem for the Navier-Stokes equations*, Comm. Pure Appl. Math., Vol. 34, 1981, p.p. 359 - 392.

[B2] Beale J. Th., *Large-time regularity of viscous surface waves*, Arch. Rat. Mech. Anal., Vol. 84, No 4, 1984, p. 307 - 352.

[G] Günter N. M., *Potential theory and its application to basic problems of mathematical physics*, Moscow, 1953, 415p.

[LS] Lagunova M. V., Solonnikov V. A., *Nonstationary problem of thermocapillary convection*, Preprint LOMI E-13-89, 1989, p. 3 - 28, SAACM Vol.1, 1991, p. 47 - 72.

[MS1] Mogilevskii I. Sh., Solonnikov V. A., *Solvability of a non-coercive initial-boundary value problem for the Stokes system in Holder classes of functions (the case of half-space)* (in Russian), Zeitschrift für Analysis und ihre Anwedungen, Bd 8(4), 1989, p. 329 - 347.

[MS2] Mogilevskij I. Sh., Solonnikov V. A., *On the solvability of a free boundary problem for the Navier-Stokes equations in the Holder space of functions*. Nonlinear analysis, a tribute in honour of G. Prodi Pisa, 1991, p. 257 - 271.

[S1] Solonnikov V. A., *Estimates of solutions of a non-stationary linearized Navier-Stokes system of equations*, Proc. V. A. Steklov Math. Inst., Vol. 70, 1964, p. 213 - 317.

[S2] Solonnikov V. A., *On differentiability properties of the solution of the first boundary value problem for nonstationary Navier-Stokes system of equations*, Proc. V. A. Steklov Math. Inst., Vol 73, 1964, p. 221 - 291.

[S3] Solonnikov V. A., *Estimates of solutions of a nonstationary Navier-Stokes system*, Zap. nauchn. semin. LOMI, Vol 38, 1973, p. 153 - 231.

[S4] Solonnikov V. A., *Estimates of the solution of an initial-boundary value problem for linear nonstationary Navier-Stokes system of equations*, Zap. nauchn. semin. LOMI, Vol 59, 1976, p. 178 - 254.

[S5] Solonnikov V. A., *On the solvability of the second initial-boundary value problem for linear nonstationary Navier-Stokes system of equations*, Zap. nauchn. semin. LOMI, Vol 69, 1977, p. 200 - 218.

[S6] Solonnikov V. A., *Solvability of the problem on the motion of viscous incompressible liquid bounded by a free surface*, Izvestia AN SSSR, ser. mathem., Vol. 41, 1977, p. 1388 - 1424.

[S7] Solonnikov V. A., *Solvability of the problem on the motion of an isolated volume of a viscous incompressible capillary liquid*, Zap. nauchn. semin. LOMI, Vol. 140, 1984, p. 179 - 186.

[S8] Solonnikov V. A., *On a non-steady motion of a finite liquid mass bounded by a free surface*, Zap. nauchn. semin. LOMI, Vol. 152, 1986, p. 137 - 157.

[S9] Solonnikov V. A., *On the evolution of an isolated volume of a viscous incompressible capillary linquid for large values of time*, Vestnik Leningr. Univ., ser, 1, 1987, No 3, p. 49 -55

[S10] Solonnikov V. A., *On a non steady motion of an isolated volume of viscous incompressible fluid*, Isvestia AN SSSR, ser. mathem., Vol 51, 1987, p. 1065 - 1087.

[S11] Solonnikov V. A., *On a non steady motion of a finite isolated mass of a self-gravitating fluid*, Algebra and analysis, 1989, Vol. 1, p. 207 - 245.

[S12] Solonnikov V. A., *On an initial-boundary value problem for the Stokes system arising in investigation of a free boundary problem*, Proc. V. A. Steklov Math. Inst., Vol 188, 1990, p. 150 - 188.

[T1] Teramoto Y., *The initial value problem for viscous incompressible flow down an inclined plane*, Hiroshima Math. Journ., Vol 15, 1985, p. 619 - 643.

[T2] Teramoto Y., *On the Navier-Stokes flow down an inclined plane*, Preprint.

## Series on Advances in Mathematics for Applied Sciences

### Aims and Scope

This Series reports on new developments in mathematical research relating to methods, qualitative analysis, interaction with computer science and mathematical modelling in the applied and the technological sciences.

This Series includes Books, Lecture Notes, Proceedings, Collections of research and review papers. Proceedings and Monographic Collections will generally have a Guest Editorial Board.

High quality, novelty of the content and potential for the applications to modern problems in applied science will be the guidelines for the selection of the content of this series.

### Instructions for Authors

Submission of proposals should be addressed to the Editor-in-Charge or to any member of the Editorial Board. Acceptance of books and Lecture Notes will generally be based upon the description of the general content and scope of the work as well as upon samples (about one third) of the book or Lecture Notes including the parts judged more significantly by the Authors themselves.

Acceptance of Proceedings will be based upon relevance of the topics and of the lecturers contributing to the volume.

Authors are urged, in order to avoid re-typing, not to begin the final preparation of the text before having received the publisher's guidelines. They will receive from World Scientific the instructions to prepare the final manuscript in a camera-ready form.

www.ingramcontent.com/pod-product-compliance
Lightning Source LLC
Chambersburg PA
CBHW050641190326
41458CB00008B/2370
*9789810208462*